ADVANCES IN FABRICATION TECHNOLOGIES FOR POWER REACTOR FUELS

The following States are Members of the International Atomic Energy Agency:

AFGHANISTAN
ALBANIA
ALGERIA
ANGOLA
ANTIGUA AND BARBUDA
ARGENTINA
ARMENIA
AUSTRALIA
AUSTRIA
AZERBAIJAN
BAHAMAS, THE
BAHRAIN
BANGLADESH
BARBADOS
BELARUS
BELGIUM
BELIZE
BENIN
BOLIVIA, PLURINATIONAL STATE OF
BOSNIA AND HERZEGOVINA
BOTSWANA
BRAZIL
BRUNEI DARUSSALAM
BULGARIA
BURKINA FASO
BURUNDI
CABO VERDE
CAMBODIA
CAMEROON
CANADA
CENTRAL AFRICAN REPUBLIC
CHAD
CHILE
CHINA
COLOMBIA
COMOROS
CONGO
COOK ISLANDS
COSTA RICA
CÔTE D'IVOIRE
CROATIA
CUBA
CYPRUS
CZECH REPUBLIC
DEMOCRATIC REPUBLIC OF THE CONGO
DENMARK
DJIBOUTI
DOMINICA
DOMINICAN REPUBLIC
ECUADOR
EGYPT
EL SALVADOR
ERITREA
ESTONIA
ESWATINI
ETHIOPIA
FIJI
FINLAND
FRANCE
GABON
GAMBIA, THE
GEORGIA
GERMANY
GHANA
GREECE
GRENADA
GUATEMALA
GUINEA
GUYANA
HAITI
HOLY SEE
HONDURAS
HUNGARY
ICELAND
INDIA
INDONESIA
IRAN, ISLAMIC REPUBLIC OF
IRAQ
IRELAND
ISRAEL
ITALY
JAMAICA
JAPAN
JORDAN
KAZAKHSTAN
KENYA
KOREA, REPUBLIC OF
KUWAIT
KYRGYZSTAN
LAO PEOPLE'S DEMOCRATIC REPUBLIC
LATVIA
LEBANON
LESOTHO
LIBERIA
LIBYA
LIECHTENSTEIN
LITHUANIA
LUXEMBOURG
MADAGASCAR
MALAWI
MALAYSIA
MALI
MALTA
MARSHALL ISLANDS
MAURITANIA
MAURITIUS
MEXICO
MONACO
MONGOLIA
MONTENEGRO
MOROCCO
MOZAMBIQUE
MYANMAR
NAMIBIA
NEPAL
NETHERLANDS, KINGDOM OF THE
NEW ZEALAND
NICARAGUA
NIGER
NIGERIA
NORTH MACEDONIA
NORWAY
OMAN
PAKISTAN
PALAU
PANAMA
PAPUA NEW GUINEA
PARAGUAY
PERU
PHILIPPINES
POLAND
PORTUGAL
QATAR
REPUBLIC OF MOLDOVA
ROMANIA
RUSSIAN FEDERATION
RWANDA
SAINT KITTS AND NEVIS
SAINT LUCIA
SAINT VINCENT AND THE GRENADINES
SAMOA
SAN MARINO
SAUDI ARABIA
SENEGAL
SERBIA
SEYCHELLES
SIERRA LEONE
SINGAPORE
SLOVAKIA
SLOVENIA
SOMALIA
SOUTH AFRICA
SPAIN
SRI LANKA
SUDAN
SWEDEN
SWITZERLAND
SYRIAN ARAB REPUBLIC
TAJIKISTAN
THAILAND
TOGO
TONGA
TRINIDAD AND TOBAGO
TUNISIA
TÜRKİYE
TURKMENISTAN
UGANDA
UKRAINE
UNITED ARAB EMIRATES
UNITED KINGDOM OF GREAT BRITAIN AND NORTHERN IRELAND
UNITED REPUBLIC OF TANZANIA
UNITED STATES OF AMERICA
URUGUAY
UZBEKISTAN
VANUATU
VENEZUELA, BOLIVARIAN REPUBLIC OF
VIET NAM
YEMEN
ZAMBIA
ZIMBABWE

The Agency's Statute was approved on 23 October 1956 by the Conference on the Statute of the IAEA held at United Nations Headquarters, New York; it entered into force on 29 July 1957. The Headquarters of the Agency are situated in Vienna. Its principal objective is "to accelerate and enlarge the contribution of atomic energy to peace, health and prosperity throughout the world".

IAEA-TECDOC-2097

ADVANCES IN FABRICATION TECHNOLOGIES FOR POWER REACTOR FUELS

PROCEEDINGS OF TECHNICAL MEETINGS

INTERNATIONAL ATOMIC ENERGY AGENCY
VIENNA, 2025

For further information on this publication, please contact:

Nuclear Fuel Cycle and Materials Section
International Atomic Energy Agency
Vienna International Centre
PO Box 100
1400 Vienna, Austria
Email: Official.Mail@iaea.org

Printed by the IAEA in Austria
August 2025

IAEA Library Cataloguing in Publication Data

Names: International Atomic Energy Agency.
Title: Advances in fabrication technologies for power reactor fuels : proceedings of technical meetings in 2021 and 2023 / International Atomic Energy Agency.
Description: Vienna : International Atomic Energy Agency, 2025. | Series: IAEA TECDOC series, ISSN 1011-4289 ; no. 2097 | Includes bibliographical references.
Identifiers: IAEAL 25-01777 | ISBN 978-92-0-117025-5 (paperback : alk. paper) | ISBN 978-92-0-117125-2 (pdf)
Subjects: LCSH: Nuclear fuels — Technological innovations. | Nuclear reactors — Materials. | Nuclear industry — Safety measures. | Nuclear power plants.

FOREWORD

The IAEA has long supported Member States in improving fuel fabrication technologies to produce reliable fuels for the operation of power reactors. Since the 1980s, the IAEA has periodically organized scientific forums, such as technical meetings and workshops, to foster the exchange of information on advances in fabrication technologies for power reactor fuels and to develop relevant IAEA publications that serve as reference materials for Member States aiming to enhance the safety, efficiency and sustainability of their fabrication programmes.

Over the past decade, there has been remarkable progress in fuel fabrication technologies, contributing to improvements in the reliability, economics, and safety of nuclear power plant fuels. Key advancements include the automation and streamlining of fabrication processes, improvements in quality assurance and quality control procedures, and the integration of cutting edge technologies such as additive manufacturing and artificial intelligence.

In the coming decade, continuing advancements in nuclear fuel fabrication technologies will be necessary to support the near term deployment of advanced fuels for both large nuclear power plants and small modular reactors, including Generation III, III+ and IV reactors (e.g. high assay low enriched uranium fuels, accident tolerant and advanced technology fuels, tristructural isotropic fuels, fuels for multiple recycling).

In light of this need for continued innovation, there is renewed interest in exchanging information on recent advancements in fuel fabrication technologies among Member States. In response to Member State requests, the IAEA organized two technical meetings to foster this exchange: the Technical Meeting on the Technical Challenges and Advances in Fuel Fabrication for Water Reactors: Recent Experiences and Future Prospects, in 2021, and the Technical Meeting on Advances in Nuclear Fuel Fabrication Technologies for Power Reactors, in 2023. This publication summarizes the key insights from these meetings. It is intended to provide useful information to Member States that are currently involved in fuel fabrication, while also serving as a comprehensive introduction to trends in nuclear fuel fabrication technologies for those who are not yet involved in this field.

The IAEA is grateful to all who contributed to this publication. The IAEA officers responsible for this publication were Y. Kultayev, K. Sim and A. Na of the Division of Nuclear Fuel Cycle and Waste Technology.

CONTENTS

1. INTRODUCTION

1.1. BACKGROUND

Nuclear fuel is a sophisticated engineering product designed for safe and reliable use in commercial nuclear power plants (NPPs). The development and implementation of new fuel fabrication technologies are essential to ensuring safe and cost effective operation of existing NPPs, as a significant percentage of fuel failures are attributed to manufacturing related issues [1]. Furthermore, the safe, economical, and timely introduction of new innovative reactor concepts, such as small modular reactors (SMRs) is unattainable without advancements in fuel fabrication technologies.

In the past decade, fuel fabrication technologies have witnessed remarkable progress within the fuel community, with a focus on enhancing the reliability, economics and safety of fuels used in operating NPPs. These advancements encompass a range of innovations, including the implementation of automated processes, enhancements in quality assurance (QA)/quality control (QC) and the adoption of cutting-edge technologies such as additive manufacturing (AM) and artificial intelligence (AI).

Over the next decade, further advancements in various aspects are anticipated, such as the introduction of high assay low enriched uranium (HALEU) fuels, the development and deployment of advanced technology and accident tolerant fuels (ATFs), and the adoption of advanced fuel cycles such as reprocessed uranium and mixed U-Pu oxide (MOX) fuels, new technologies like computer-aided technology, AM or 3D printing, and AI are also expected. Moreover, there will be a focus on the development and/or deploying fuels for SMRs, including light water reactor (LWR), fast neutron reactor types, coated particle fuels for modular gas cooled reactors, and fuels for molten salt reactors.

Recognizing the importance of these advancements in fuel technologies and the renewed interest among Member States in those advancements, in 2021 and 2023, two Technical Meetings on the "Technical Challenges and Advances in Fuel Fabrication for Water Reactors: Recent Experiences and Future Prospects" and on the "Advances in Nuclear Fuel Fabrication Technologies for Power Reactors" were organized by the IAEA, with the objective of providing a platform for Member States to facilitate the exchange of information on nuclear fuel technology advancements for power reactors.

The technical meeting held in 2021 comprised five technical sessions on i) powder and pelletizing, ii) fuel rod manufacturing and inspection, iii) fuel assembly and appendage welding, iv) fuel pellet examinations, and v) building fuel technology in nuclear embarking countries.

The technical meeting held in 2023 was organized with five technical sessions on i) powder and reconversion, ii) sintering, pelletizing, advanced pellet fabrication, iii) cladding, fuel rod/assembly components, iv) advancements in fuel fabrication facilities and inspection, and v) fabrication technologies for new fuel development.

The technical meeting held in 2021 was attended by 39 participants from 22 Member States, delivering 14 oral presentations. The attending countries and organizations were Argentina, Bangladesh, Belarus, Brazil, Canada, China, Egypt, France, Hungary, India, Indonesia, Iran, Jordan, Korea, Pakistan, Romania, Russian Federation, Sweden, Switzerland, Türkiye, Ukraine, and United Arab Emirates.

The technical meeting held in 2023 was attended by 48 participants from 16 Member States, delivering 15 oral presentations. The attending countries and organizations were Argentina, Armenia, Brazil, Bulgaria, Canada, China, France, India, Iran, Korea, Pakistan, Romania, Russian Federation, South Africa, Spain, and United Arab Emirates.

Considering the relevance of the technical contributions provided during these technical meetings the development of a TECDOC summarizing the materials presented was identified as being a highly beneficial follow-up action to these technical meetings, reflecting the outputs from the IAEA efforts in promoting nuclear fuel technologies for operating and advanced nuclear reactors.

1.2. OBJECTIVE

The current TECDOC aims at capturing the status in technological development in fuel fabrication across various technological stages, including powder and reconversion, sintering, pelletizing, cladding, fuel rod and assembly components. It will also address advancements in streamlining fabrication processes and inspection, along with experiences and future prospects in fuel fabrication.

In addition, this publication is also intended to provide Member States with a high level reference on the status and trend in fuel fabrication technology developments, with supporting evidence described in the proceedings, session summaries and conclusions of the Technical Meetings, including the extended abstracts and full papers presented at the meetings.

1.3. SCOPE

This TECDOC presents the proceedings of the above mentioned technical meetings held in 2021 and 2023. It compiles summaries of the technical sessions and conclusions drawn from group discussions, along with extended abstracts and full papers submitted and presented during the events.

This TECDOC does not cover the topics related to the fabrication of fuel with reprocessed uranium or HALEU regardless of their high topic interest in some countries. None has addressed these topics in both technical meetings.

1.4. STRUCTURE

Section 1 provides the introduction to the publication with the background, objective, scope, and structure of the publication. Section 2 summarizes the five technical sessions and includes observations for each subject of the technical sessions of technical meetings providing insights and future prospective. Section 3 highlights the main summary and conclusions from the meetings. Extended abstracts and full papers of the presentations delivered at the technical meetings are also included in this publication.

2. SUMMARY OF TECHNICAL SESSIONS

2.1. TECHNICAL SESSION I: POWDER AND RECONVERSION

Prepared by Mr Jae Ho Yang (Republic of Korea, Korea Atomic Energy Research Institute)

The properties of nuclear fuel powder are crucial in determining the quality of nuclear fuel. Consequently, the reconversion and handling of nuclear fuel powder are subjects of extensive research and development (R&D). Despite having identical chemical compositions, nuclear fuel powders produced through different production routes exhibit diverse properties and behaviours. Furthermore, each stage of fuel powder treatment such as mixing, milling, and granulation can further alter these properties and behaviours. It is crucial to understand how powder properties change with various process variables to design and optimize fabrication processes, ultimately enhancing the quality of the nuclear fuel.

Studies on reconversion focus on developing a new reconversion process or modelling of the overall process to optimize the process and improve powder properties.

With the ongoing manufacturing of nuclear fuels, advancements in the reconversion to recycle the fuel scraps are necessary. While the ammonium uranyl carbonate (AUC) process can produce free-flowing UO_2 powder, its high solubility in water necessitates liquid waste treatment facilities. Conversely, the ammonium diuranate (ADU) process produces powder that is too fine, resulting in complexities in powder handling processes. Therefore, there is a demand for research aimed at developing a new wet reprocessing process that can address the drawbacks of existing wet processes.

The dry conversion (DC) process is the most widely utilized reconversion method. Due to the highly complex chemical reactions occurring in the reactor, research is underway to develop numerical models of the entire process aimed at enhancing reactor performance in both powder quality and production throughput. Such models also serve as valuable tools for educating production/maintenance personnel and engineers.

The primary objective of R&D in powder processes is to enhance existing processes by gaining a better understanding of process operations through coupled experimental and modelling studies. The ultimate goal is to build a multidimension, multiphysics, and multiscale simulation tool for all stages using predictive models.

This session highlights advances in the reconversion processes, as well as modelling and optimization of powder processes, with a total of 4 presentations: one from the 2021 and 3 from the 2023 technical meetings.

The first presentation (ID#1) described a novel wet reconversion process – ammonium uranate hydrate (AUH) process developed by KEPCO Nuclear Fuel (KEPCO NF) in Republic of Korea to produce high-quality fuel-grade UO_2 powder from scrap materials. The second presentation (ID#2) introduced a project of Framatome to model the complex chemical solid–gas reactions in the reactor and calciner during the integrated dry route process. The third presentation (ID#3) focused on research activities on the rheological behaviour of UO_2 powders and granules during the ceramic process at CEA (Commissariat à l'Énergie Atomique et aux Énergies Alternatives) in France. The last presentation (ID#4) provided an overview of the MIMAS (MIcronization of a MASter blend) process, a manufacturing process utilized to produce MOX pellets at Orano in France.

2.1.1. Summary of presentations

ID#1 Republic of Korea (KEPCO NF), Development of ammonium uranate hydrate (AUH) wet reconversion process for the production of nuclear-grade UO_2 powder from uranyl nitrate hexahydrate solution – Mr Seungchul Yang

The presentation focused on the AUH process which is a novel wet reconversion process developed by KEPCO NF to produce high-quality fuel-grade UO_2 powder from the scraps. It provided a comprehensive overview of the significant findings and experiences during the development phase for the AUH conversion process, highlighting the properties of UO_2 powders produced by the AUH process and those of the UO_2 test pellets manufactured from these powders.

The current wet reconversion processes used to recover scraps produced during nuclear fuel manufacturing are intricate, involving multiple unit operations and extensive wastewater treatment. KEPCO NF has sought to streamline the recovery of high-quality, fuel-grade UO_2 powder. A new wet reconversion process for uranyl nitrate hexahydrate solution has been developed, featuring an innovative pulsed fluidized bed reactor. This study evaluated the chemical properties of the intermediate AUH product and the resulting UO_2 powder, highlighting the advantages of this method compared to existing wet processes.

The UO_2 powder produced through the proposed process has improved fuel pellet properties compared to those obtained from conventional wet conversion methods. Performance tests indicated that the resulting UO_2 powder meets critical requirements for fuel pellets, including sintered density, enhanced re-sintered density, and optimal grain size. This approach offered a streamlined manufacturing process for nuclear-grade UO_2 powders suitable for use in UO_2 nuclear fuel pellets. This presentation has concluded with important results obtained and experiences gained from the developmental works related to the AUH reconversion process.

ID#2 France (Framatome), Modelling of the dry conversion process – Mr Jeremy Bischoff

Framatome introduced a project to model complex chemical solid–gas reactions that occur in the reactor and calciner during the DC process. This modelling endeavour has offered several advantages, such as performing parametric studies without the use of production equipment, optimizing processes for better performance, and providing training tools. Therefore, the modelling of this process is expected to enable a better understanding of the DC process.

The conversion of UF_6 to UO_2 at the Framatome fuel manufacturing plant has been achieved through the DC process, which eliminates the generation of liquid contaminated wastes and enhances productivity. This complex process involves chemical solid–gas reactions, making it challenging to fully understand all reactions in the reactor and calciner. To address this, Framatome initiated a project to model the process. The modelling approach has progressed step by step, incorporating new reactions or functions at each stage. Initially, the geometry of the conversion reactor was established, followed by modelling gas flows and focusing on chemical reactions at the gas injector, then incorporating the influence of temperature, and finally the precipitation of UO_2F_2 particles. The model's validation was achieved using experimental data collected during the reaction. Parametric sensitivity analyses were conducted to assess the impact of different parameters on the process and optimize reactions. For instance, by varying the N_2 flow rate, the impact on the distance from the injector where UO_2F_2 particles are created could be evaluated, helping determine an appropriate flow rate range to prevent particle accumulation on the injector and ensure particles do not form on the reactor walls. The utilization of modelling techniques has provided better control of the process.

ID#3 France (CEA), Uranium oxide powder studies for nuclear fuel fabrication – Ms Anne-Charlotte Robisson

The presentation provided an overview of CEA's research activities on the rheological behaviour of UO_2 powders and granules. A comprehensive understanding of the rheological properties of powders is crucial for ensuring reliable nuclear fuel fabrication and accommodating various feeding powders. The primary objective of this work is to gain deeper insights into the flowability of UO_2 powders to predict the overall behaviour of these granular materials in processes, thereby improving fabrication processes and facilitating the transition from laboratory to industrial scale.

To accomplish this goal, the CEA has established a dedicated experimental research programme focused on the rheological properties of UO_2 powders. Previous studies primarily involved developing empirical models correlating flowability with particle characteristics. However, current research emphasized shape factors, particle structure, and surface properties to predict flowability in different scenarios. The research programme was structured around three main research axes: the description of powder agglomerates, the study of the agglomerate behaviour, and the prediction of powder flowability. This prediction was performed using a variety of methods: van der Waals force calculation, statistical approaches including linear regression models and correlation matrices, and device instrumentation. Furthermore, this work was part of a comprehensive approach that includes numerical simulations of fabrication processes.

ID#4 France (Orano), MOX fuel process: Powders and pelletizing fabrication experience – Mr Jean-Michel Marin

This presentation introduced the MIMAS process, a manufacturing process for MOX pellets at Orano in France. The MIMAS process is a two-step process using a master blending powder that allows for the adaption of various powders produced through wet and dry processes. Noteworthy benefits of MIMAS process include high recycling rates and the ability to have an extended range of plutonium content. Moreover, it has demonstrated its effectiveness through the commercial production of thousands of tons of produced material.

The MIMAS process consists of six-unit processes, starting with the preparation of primary blend powder containing UO_2 and PuO_2, which is then mixed again with UO_2 powder to produce secondary blend powder. The first step involves homogenizing the primary blend, where four types of powder (UO_2, PuO_2, scrap, and additives) are blended before milling. Subsequently, the primary powder is milled in the second step. Then, in the third step, the primary powder is prepared into uniform sized particle powder through force sieving. Following this, in the fourth step, the primary powder is mixed with UO_2 and poreformer in the secondary blend homogenization step. Then, in the fifth step, the secondary blend is combined with lubricant, and in the final step, the secondary blend is pelletized into green pellets.

2.1.2. Observations: Insights and Future prospective

The demand for new technologies in the reconversion process remains notable. Despite the limited research on innovative reconversion processes, these technical meetings reaffirmed the continuous research interest in advanced reconversion processes aimed at recycling uranium scrap resources generated during the fuel manufacturing process or improving the process through modelling and experimental testing.

The importance of strong iterations between experiments and modelling was emphasized, underscoring the pivotal role of coupling experimental tests with modelling in process development. The utilization of modelling based on experimental data enhances understanding of powder behaviour throughout the process, enabling efficient optimization or scale-up. Furthermore, modelling offers several advantages, including the ability to conduct parametric studies without the need for production equipment, optimize processes to enhance performance, and serve as training a tool.

The importance of R&D on reconversion and powder is expected to continue increasing. In addition to the topics addressed at these two meetings, it is possible to identify major R&D trends and prospects in the areas of reconversion and powder production.

Here are selected research trends where further study on reconversion and powder processing is needed:

- Use of non-traditional dopants in ATF.
- Application of advanced manufacturing methods such as AM.
- Extended use of diverse burnable absorbers for LEU+/Boron-free core.
- Developing nuclear fuel technologies for advanced reactors including SMR:
  - Direct reconversion from UF_6 to non-oxide fuels;
  - New powder preparation processes such as milling and blending;
  - Recycling of new fuel scraps.

2.2. TECHNICAL SESSION II: SINTERING, PELLETIZING, ADVANCED FUEL PELLET FABRICATION

Prepared by Ms Anne-Charlotte Robisson (France, Commissariat à l'Énergie Atomique et aux Énergies Alternatives)

Nuclear fuel fabrication, whether for LWRs or future SMRs, continues to be the subject of significant R&D efforts. Ongoing studies on sintering, pelletizing, and advanced fuel fabrication are centred on optimizing processes, improving fuel properties for reactor behaviour, and improving characterization methods. These studies being conducted have different objectives.

One of the objectives is to optimize processes, in particular, to limit manufacturing rejects, increase production rates, limit the production of contaminating dust, limit worker exposure or adapt processes to a certain variability in raw materials. In this context, R&D works mentioned in the literature are aimed at improving existing processes through a better understanding of process operations, often by means of coupled experimental and modelling studies. This understanding is expected to lead to improvements in the three main stages of the fabrication processes: powder mixing, pressing, and sintering.

Another objective is to improve fuel properties to enhance reactor performance. Research is underway to explore innovative fuels such as ATF and full flex fuels. These new fuels are leading to an upsurge in design and manufacturing work. Specifically, the fuel compositions or microstructures are optimized to achieve characteristics that lead to increased safety margins: increased thermal conductivity, homogeneity, density, and fission product retention. This leads to the development of different types of fuel pellet concepts: coarse-grained microstructures with or without dopants, fuel containing a neutron absorber, ring pellets, composite fuel, and microcell structured fuel. Studies therefore focus on exploring the various possible experimental ways to obtain these different types of fuel. Subsequent irradiation experiments are generally conducted to validate the in-pile behaviour of these new fuels. The technologies developed have potential applications in LWR and SMR fuels.

In relation to these objectives, the organizations are also seeking novel characterization methods to provide a precise description of the pellets obtained. These studies will establish correlations between process steps and pellet characteristics, and between pellet characteristics and reactor behaviour.

The presentations delivered in this session are part of this research context and illustrate the aforementioned three objectives. A total of six presentations were given in this session: two were presented at the 2021 technical meeting and four at the 2023 technical meeting. Specifically, two presentations (ID#5 and ID#6) focused on UO_2 sintering, two others explored fuel improvement using additives (ID#7 and ID#8) and the remaining two addressed UO_2 pellet characterization (ID#9 and ID#10).

2.2.1. Summary of presentations

ID#5 France (Framatome), Modelling of the fuel sintering process – Ms Gaëlle Raveu

Research and development activities focused on process optimization can be illustrated by the modelling study of sintering furnaces at Framatome's UO_2 pellet production plants, presented by G. Raveu of Framatome, France. This study presented a detailed account of the sintering mechanisms that result in a reduction of intergranular porosity and an increase in grain size. Furthermore, it outlined the primary sintering stages and furnaces utilized at Framatome's facilities in France, Germany, and the USA, with a particular focus on the mechanisms that occur during the sintering of UO_2 pellets.

The objective of modelling the sintering process was to provide a theoretical description of this stage to optimize the process, compare different furnaces and conduct sensitivity studies. Models of the furnace, the boats, and the heating elements were constructed based on their geometry to achieve this goal. The boats were considered to be uniformly filled with UO_2, regardless of the geometry of the pellets. The furnace aeraulics were also modelled, considering thermal gradients and gas repartition. The resulting global model has yielded temperature values that closely align with the real values. Further studies are to continue at pellet, grain, and atomic scale, particularly by considering gas exchange between UO_2 and the atmosphere.

ID#6 Brazil (INB), Evaluation of the correlation between green and sintered density in UO_2 pellets – Mr Joao Paulo Carnaval

The sintering step has been also studied experimentally. The presentation by Mr Carnaval of Indústrias Nucleares do Brasil (INB), Brazil, investigated the correlations between green density and sintered density for doped and undoped UO_2 pellets. Various mixtures were examined based on their U_3O_8 concentration and the type and amount of dopants present (Al_2O_3, Cr_2O_3, Nb_2O_5). The results demonstrated a linear correlation between the raw density and the sintered density for UO_2 and UO_2–U_3O_8 mixtures.

Furthermore, as anticipated, an increase in the U_3O_8 content has resulted in a reduction in the sintered density. When considering doped pellets, the dopant type has a more significant impact on the sintered density than the amount of dopant or the raw density for doped pellets. Moreover, the introduction of dopants can significantly increase grain size, depending on the specific dopant utilized. In conclusion, this study highlighted the importance of green density, U_3O_8 content, and dopant type (Al, Nb, or Cr) in controlling the sintered density of UO_2 pellets. Additionally, dopants can also be utilized to increase grain size to obtain an optimal end product.

ID#7 Argentina (CNEA), Advanced fabrication and characterization of Gd pellets for SMR fuel in Argentina – Ms Maria Florencia Parrado

Additives such as Gd_2O_3 are used in the manufacture of burnable neutron absorbers. The study presented by Parrado of Comisión Nacional de Energía Atómica (CNEA), Argentina, focused on the production and characterization of 3.1% enriched UO_2–Gd_2O_3 fuel pellets for SMR fuel in Argentina. This work aimed at finding manufacturing parameters compatible with existing technologies and enabling industrial scale production in Argentina. The process under investigation utilized UO_2 and Gd_2O_3 powders and included classical mixing, pressing, and sintering steps. Microstructural characterization was employed to analyse existing phases, and determination of grain and pore distributions. At the same time, UO_2–Gd_2O_3 powder and pellet characterization systems were developed to control raw materials, and intermediate and finished products. Improvements in the homogeneity of the gadolinium distribution within UO_2 were achieved by optimizing the mixing step to obtain pellets without Gd islands. Two hundred UO_2–Gd_2O_3 pellets were manufactured for irradiation in the Halden reactor. A change of scale was performed to corroborate the initial laboratory scale results. Once the reproducibility of the results had been validated, the process qualification stage proceeded. Upon successful completion of the qualification stage, the production phase can begin.

ID#8 Republic of Korea (KAERI), Development status of high thermal conductive ATF pellet and burnable absorber fuel pellet fabrication and property evaluation of UO2–metal composite pellet – Mr Dong-Joo Kim and Mr Jae Ho Yang

In the context of improving fuel performance, the studies presented by D.J. Kim and J.H. Yang presented different types of ATF fuel currently being researched at Korea Atomic Energy Research Institute (KAERI). This section provides a synthesis of the findings from these two studies. Specifically, the authors described the results of investigations into a microcell structured composite UO_2–metal fuel that improves the thermal conductivity of UO_2 pellets, thereby reducing the temperature, and temperature gradients thus increasing the safety margin. Two different concepts were studied, depending on how the metallic phase was distributed within the UO_2 matrix: metallic microcell (continuous networks of the metallic particles) or metallic microplate (directionally oriented plates). Mo and Cr, two types of refractory metals were selected, based on their melting temperature, thermal conductivity, and absence of chemical reaction with UO_2 during manufacturing. The pellets were manufactured using a conventional ceramic sintering process, and the influence of various manufacturing parameters was analysed. Examples of microstructures were presented for both concepts and both metal particles. Thermal and irradiation properties were measured and the benefits of these concepts were demonstrated: the thermal conductivity of these fuels is 80% higher than that of a reference UO_2 fuel. The development of these new fuels is underway, with the study of their behaviour under longer term irradiation.

ID#9 Argentina (CNEA), Developments and methodology for the characterization of the mechanical integrity of PHWR fuel pellets – Mr Juan Pablo Medina

In addition, to optimize fuel microstructure and performance, R&D is also underway to develop pellet characterization methods. For example, measurement methods are being developed to study the behaviour of whole pellets, as well as methods for fine characterization of the microstructure. Mr Medina of CNEA, Argentina, presented a methodology for characterizing the mechanical integrity of PHWR (Pressurized Heavy Water Reactor) fuel pellets. Inadequate mechanical integrity of pellets can result in chips in the pellet–cladding gap or a missing pellet surface. This can subsequently lead to the formation of hot spots, stress concentrations, fission product accumulation, or a lack of support. In this context, a methodology has been developed by the CNEA to characterize pellet structural integrity. This methodology includes the use of ceramographic and visual examination techniques as well as three distinct types of tests: compression, percussion, and loading and rotation tests. Percussion tests developed revealed that the surface of the pellet on the lower die side is more prone to chipping and breakage than that on the upper die side. Loading and rotation tests offer a straightforward way of execution that improves early detection of chipping of CANadian deuterium-uranium reactor (CANDU) fuel pellets.

ID#10 Iran (AEOI), The UO_2 pellet ceramography and microstructure analysis – Mr Abdollah Riahi

The study presented by A. Riahi from the Atomic Energy Organization of Iran (AEOI) focused on the microstructural analysis of UO_2, specifically on the analyses used to reveal grain boundaries. These analyses were conducted through the application of ceramographic techniques, including sample preparation, microscopic examinations, and image analysis. The results of these analyses permit the determination of the average grain size and grain size distribution, the pore size distribution and pore content, and the pore shape histogram. Furthermore, two etching methods were investigated: thermal etching International Organization for Standardization (ISO) ISO 16793 [2] and chemical etching American Society for Testing and Materials (ASTM) ASTM C1868 [3]. The advantages and disadvantages of both methods are presented, followed by comparative analyses of UO_2 pellets. The results showed that thermal etching was of better quality. Some grains can be over-etched or under-etched in chemical etching, although thermal etching may not increase grain size. This study also presented results on pore size and shape distributions. Grain size ranged from 8 to 20 μm.

2.2.2. Observations: Insights and Future prospective

Although they do not cover the wide range of R&D topics that could be addressed in this session, these presentations highlighted the diversity of approaches leading to optimization of fabrication processes, microstructure or fuel characterization. The main optimization principles are based on:

- Simulation, modelling or empirical correlation research to optimize manufacturing conditions and perform sensitivity studies;
- The application of additives or dopants to improve fuel behaviour in terms of fission product retention, enhance thermal conductivity or incorporate neutron absorbers;
- Material characterization to obtain quantifiable data in a more reliable and reproducible manner.

The discussions were fruitful and the many technical questions helped broaden the topics and give a global overview of pellet fabrication. The different points listed below illustrate some areas of interest that have been identified:

- Manufacturing conditions: sintering conditions, mixing methods, pellet composition;
- Data used for modelling and correlation;
- Pellet characterization: quantifying the homogeneity of cation distribution in pellets, etching methods to reveal grain boundaries, quantifying and on-line methods to analyse pellet surface defects;
- Pellet requirements: in addition to classical requirements (chemical composition, isotopic composition and homogeneity, pellet surface finish, density and geometry, microstructure) data on pellet brittleness could be considered.

In addition to the subjects addressed at these two meetings, it is possible to identify major R&D trends and prospectives in fuel fabrication:

— A better understanding of the mechanisms that control the sintered density of doped pellets;
— Closer modelling of real-world conditions: variation in the quantity of pellets in the sintering boats, taking into account the gas exchange between the pellets and the atmosphere;
— Development of a method to measure and quantify uranium and dopant homogeneity in pellets;
— Development of on-line and off-line measurement methods to quantify pellet surface defects (related to the presentations in Session IV (see Section 2.4)).

2.3. TECHNICAL SESSION III: CLADDING, FUEL ROD AND FUEL ASSEMBLY COMPONENTS

Prepared by Mr Ike Francisco Dimayuga (Canada, Canadian Nuclear Laboratories)

Fuel cladding serves an important function in fuel performance – it is the primary containment of fission products and actinides resulting from the fission and neutron capture processes. The fuel cladding has to survive very challenging conditions of neutron and gamma irradiation, high temperature and high pressure coolant, and turbulent flow conditions, among others. As such, the fuel cladding of choice for nuclear power applications has to maintain good neutronic, thermal, mechanical, and corrosion properties during operation. Additionally, the cladding needs to maintain these properties to a reasonable extent even after irradiation to different burnup levels.

Currently, the cladding alloys used in most power reactors are zirconium based, such as Zircaloy-2, Zircaloy-4, and Zr–Nb alloys. Development efforts have been remarkably successful as illustrated by the introduction of E110/E110M/E125, M5, E635, Q12, ZIRLO/Optimized ZIRLO, AXIOM, HANA, etc. [4]. Such development efforts continue to look at improved alloys or materials for cladding applications, including ATF cladding such as FeCrAl or coated Zr-based alloys. The development of these new alloys or coating materials has to consider fabrication aspects to ensure that adopting a different alloy/material will not cause a drastic change in the cladding manufacturing process, as well as any downstream operations of manufacturing the fuel assembly (FA). The coating developments bring additional fabrication steps that have to be mastered to guarantee appropriate behaviour in reactor during normal and accident conditions such as adherence, deformation or neutronic absorption.

Other development efforts focus on optimizing and improving the fabrication process for fuel rods. One such area is the brazing of appendages, i.e. bearing and spacer pads, onto the cladding of CANDU fuel. Other areas of fuel rod fabrication being investigated include new joining techniques such as resistance brazing, laser welding, and laser brazing.

Fuel assembly components of CANDU fuel include the finished fuel rods (UO_2 pellets contained within the cladding with bearing and spacer pads on the external surface of the cladding, and the cladding ends welded with endcaps) and the endplates. The fuel rods are welded onto the endplates, which maintain the geometry of the FA. When fully assembled, the FA for CANDU is often referred to as a fuel bundle.

For pressurized water reactors (PWRs), FA components include the finished fuel rods (containing UO_2 and neutron absorber doped pellets, a spring to maintain the plenum space within the cladding, and endcaps welded onto the cladding ends), top and bottom nozzles (which provide mechanical support to the FA structure), grid spacers (to ensure precise spacing of the fuel rods), and the guide thimble tube (an empty tube for control rods or in-core instrumentation). Boiling water reactor (BWR) FAs are similar to those in PWRs, with one major exception: BWR FAs have a duct surrounding the fuel rods to prevent radial vapor drift. Typically, four FAs are arranged in a square module with a cruciform control blade in the centre.

This session comprises several presentations focusing on new cladding alloy development and improvements in fuel rod assembly operations. Two presentations (ID#11 and ID#12) discussed R&D efforts to eliminate toxic beryllium (Be) as a filler metal during the brazing of appendages onto CANDU cladding. One presentation (ID#11) examined the use of an alternative joining process, i.e. resistance welding as opposed to brazing, to eliminate Be. The approach to resistance welding of appendages was adopted from the manufacture of FAs intended for PWRs. This presentation showcased the results of initial trials involving some mechanical tests and microstructural examination. The other presentation (ID#12) explored the use of a different filler metal while maintaining the same induction brazing process. This presentation highlighted the results of a multi-year programme to replace Be with stainless steel 316 (SS316). Several out-reactor and in-reactor tests demonstrate the suitability of using SS316 as an alternative filler metal for induction brazing of appendages onto CANDU fuel cladding.

New cladding alloy development was the subject of three presentations(ID#12, ID#13, and ID#14), and one presentation (ID#14) dealt with the work on coated Zr-4 cladding as a contribution towards having ATF. Another presentation (ID#13) discussed the development of an entirely new Zr alloy, N36, and presented results of out-reactor testing of specimens and irradiation testing of lead test assemblies, while the other presented (ID#12) results of testing and characterizing a Zr–Nb alloy, with special focus on the effects of brazing and welding on mechanical and corrosion properties. A summary of each presentation is provided in the following sections below.

2.3.1. Summary of presentations

ID#11 Argentina (CNEA), Alternatives to replace the use of beryllium for appendage attachment on CANDU fuel elements – Mr Alejandro Bussolini

This presentation discussed the work done at CNEA to eliminate the use of beryllium as a filler metal for brazing of appendages onto the cladding of CANDU fuel for use in Argentina's CANDU-6 reactor — Embalse. This work demonstrated the benefit of adopting technologies used in other operating PHWR reactors in Argentina (Atucha 1 and 2) and applying them to Embalse in terms of fuel fabrication. While there are some differences in FA components between the two types of PHWRs, such as overall length, there are also some similarities such as the concentric arrangement of 37 fuel elements. The authors have noted that the appendages in Atucha fuel rods were attached by resistance welding. Despite the difference in wall thickness of the cladding between the Atucha fuel and the Embalse fuel (with the latter having thin wall, collapsible cladding), they have conducted tests using resistance welding of appendages onto the Embalse fuel. Although some preliminary tests showed good results, especially in terms of microstructure, some issues still need resolution before this technology can be used as the standard process for fuel fabrication.

ID#12 Canada (CNL), Developments in fuel fabrication at CNL – Mr Francisco (Ike) Dimayuga

This presentation provided an overview of the development work on fuel fabrication technologies at the Canadian Nuclear Laboratories (CNL) with topics ranging from the study of new filler materials for brazing to the development of alternative joining techniques applied to zirconium alloys. Results of the out-reactor and in-reactor testing of stainless steel as a braze filler material to replace beryllium were discussed, with the overall conclusion that SS316 is a suitable replacement braze filler material to Be, with no performance-related issues. Additionally, preliminary results of exploratory work on laser welding and resistance brazing as alternative joining techniques for fuel fabrication were presented. Some of CNL's work on advanced fuel cladding alloys such as Zr–Nb, was also discussed, particularly in terms of results of out-reactor testing for mechanical and corrosion properties that can inform their potential suitability to high burnup or ATF applications. It was observed that before being adopted for use in PHWRs, there is a need to develop alternative joining techniques to minimize the detrimental effects of current welding/brazing techniques on Zr–Nb. The future direction of the R&D effort was provided, including the joining of FeCrAl and coated Zr-4.

ID#13 China (NPIC), Industrial development of N36 zirconium alloy cladding tube – Mr Chunrong Xu

This presentation discussed the efforts of the Nuclear Power Institute of China (NPIC) in developing an industrial process to produce a new Zr alloy N36 (Zr–1.0%Sn–1%Nb–0.3%Fe), including process exploration, process verification, process scaling, process optimization, process standardization, and industrialization. The N36 alloy cladding tubes exhibit significantly better performance in terms of out-of-pile corrosion resistance, hydrogen absorption resistance, creep resistance, fatigue resistance, iodine-induced stress corrosion resistance, and high-temperature oxidation resistance compared to the Zr-4 alloy. Furthermore, lead FAs were prepared using N36 zirconium alloy cladding tubes, and they successfully underwent irradiation tests in commercial reactors to very high burnups of 55 GWd/tU. The post-irradiation examination of the fuel rods confirmed the excellent in-reactor performance of N36 zirconium alloy. Through this research, comprehensive knowledge of the industrial manufacturing process and application technology of N36 zirconium alloy cladding materials were achieved, and data on the in-pile and out-of-pile performance of N36 zirconium alloy were obtained, thereby realizing the goal of engineering application.

ID#14 Republic of Korea (KAERI), Development Status of ATF and 3D printing technology for nuclear fuel – Mr Hyun-Gil Kim

This presentation discussed KAERI's R&D activities on the development of ATF, with their focus on coating technologies both in terms of the coating materials, e.g. Cr, CrAl, FeCrAl, and the coating techniques, such as arc ion plating (AIP) and 3D laser coating. The behaviour of coated cladding materials was evaluated through out-of-pile and in-pile tests. In particular, oxidation resistance in a high temperature oxidizing atmosphere is improved by Cr, FeCrAl and FeCrAl alloy coatings. As a result of in-pile test in research reactor, cladding coated with CrAl and FeCrAl alloys showed that the integrity of the coating layer was well maintained. Based on these performance evaluation results, it was analysed that the safety of the surface-modified cladding was improved compared to the commercial cladding. Additionally, work on 3D printing technology as applied to fuel is on-going. Aspects of this technology are being explored, including modelling, manufacturing, sensing, and QC.

2.3.2. Observations: Insights and Future Prospective

It can be observed that a significant fraction of the R&D effort for CANDU fuel is focused on eliminating Be in fuel fabrication. This can be achieved through various means, one of which is to replace Be with an alternative material that is more environmentally and human health compatible. Another approach is to use a different joining technology altogether. There are pros and cons of each approach. One pro of using an alternative filler material, e.g. SS316, is that while the filler material is different, this minimizes any changes in the fuel manufacturing process. The advantage of using an alternative joining process is that it presents an opportunity to leverage technologies from other fuel fabrication processes that may already have significant operating experience to draw on, as seen in the example of adopting resistance welding to replace induction brazing to attach appendages. Both approaches still need to go through an extensive testing programme, including out-reactor and in-reactor testing, as well as demonstration of compatibility with full production scale fuel manufacturing.

Future areas of work in this area may take the form of further development of alternative joining techniques in addition to resistance welding, such as resistance brazing, laser welding, electron beam welding, and laser brazing. The technical merit of using SS316 as an alternative filler material for induction brazing appears to have been proven; however, it still requires demonstration irradiation of several fuel bundles produced at scale to further validate this technology.

Development work on improved cladding alloys, including Zr–Nb and N36, is advancing cladding applications for higher burnups and ATF. This ongoing trend, which has been developing for several decades, is driven by the need to improve fuel performance and reliability, as well as by the desire to establish a reliable domestic supply chain of these materials. Further research work will continue to explore aspects of joining these new alloys, particularly regarding the ability to use the same joining techniques employed for standard Zr-4 cladding.

Development in ATF applications has also seen the use of engineered coatings on standard Zr-based alloys. This trend can provide more significant gains in cladding development because by only modifying the external surface of the cladding, the industry can leverage the vast irradiation experience of the base cladding material. This could potentially reduce the effort and timelines involved in developing, qualifying, licensing, and implementing this technology. Another interesting aspect of coating the cladding surface is the prospect of developing complex, multilayered structures, leading to a product with specific surface functionalities. This is an interesting prospect for the future to improve the corrosion, mechanical, and thermal-hydraulic behaviour of fuel rods in a reactor with almost limitless possibilities. It is expected that as more novel and improved coating technologies are demonstrated, including additive manufacturing and other advanced coating techniques, their contribution towards improvement in cladding materials will also increase significantly.

2.4. TECHNICAL SESSION IV: ADVANCEMENTS IN FABRICATION FACILITIES AND INSPECTION

Prepared by Mr Jeremy Bischoff (France, Framatome)

Nuclear fuel manufacturing facilities worldwide are continuously improving to adapt to the fabrication of new fuel designs, the increase in safety standards, the demand for higher quality products, and the reduction of fabrication costs in a competitive environment. For example, new fuel types such as ATF or SMR fuels are being developed with different designs, which impact manufacturing. In the same way, the Russian fuel manufacturer TVEL has built the capacity to fabricate western PWR designs, and inversely, western manufacturers are starting to build Water Water Energy Reactor (WWER) fuel capacities [5, 6]. This incentivizes fuel manufacturers to increase operational flexibility to manufacture various designs, prompting process innovation and improvements.

The global trend to reduce fuel failures during reactor operation has also driven manufacturing facilities to enhance production quality since a small proportion of fuel failures are linked to manufacturing issues [1] but with usually undesired impacts on reactor operation. In particular, additional QC systems were installed or optimized to prevent defective products from being sent to power production facilities. To reduce costs of inferior quality, fuel manufacturers therefore had to improve the whole manufacturing process and equipment.

Furthermore, safety standards are continuously enhanced to limit operational dose and mitigate the likelihood of accidents, especially given the current trend of increasing fuel enrichments beyond 5%. This often creates significant constraints on production equipment and processes, necessitating adaptations. This also drives the development of more automatic processes and specifically, automatic control and inspection systems to reduce operational dose as well as to reduce human factor effects on production quality with an overall increase in performance.

Additionally, following the COVID-19 pandemic and other geopolitical transformations, a prevailing inclination towards heightened national sovereignty has spurred certain countries to prioritize localisation and the expansion of their national production capacity to limit potential supply issues. Consequently, this shift is expected to result in the emergence of new market entrants, thereby increasing international competition for fuel manufacturing globally.

Consequently, nuclear fuel manufacturing facilities are required to promptly adapt to the rapidly evolving environment, necessitating continuous process improvements, equipment modifications with additional operational flexibility, and often an increase in automation.

This session focused on the upgrades to certain manufacturing facilities and specific equipment, featuring a total of 6 presentations: 3 from the 2021 and 3 from the 2023 technical meeting. The first presentation (ID#15) provided an overview of the different manufacturing facilities of TVEL in Russia and the recent modifications implemented to manufacture TVS-K FAs for the western PWR market. The second presentation (ID#16) highlighted the improvements performed on the ZrB_2 fuel pellet coater in the Baotou factory in China to improve its performance and stability. The third presentation (ID#17) introduced the development of an automatic visual inspection system for CANDU tube welding preparation to ensure the machined groove meets necessary dimensional and cleanliness requirements before welding. Similarly, ENUSA in Spain has also developed new inspection equipment to improve the fuel manufacturing quality inspections with visual pellet and fuel rod surface inspections, which will be described in the fourth presentation (ID#18) along with the use of augmented reality to enhance manufacturing performance. The two final presentations (ID#19 and ID#20) described the replacement of fuel rod active gamma scanners with high speed passive scanners: one from ENUSA in Spain and one from China North Nuclear Fuel Co (CNNFC) in China.

2.4.1. Summary of presentations

ID#15 Russia (TVEL), Development trends and recent achievements in fuel fabrication technology at TVEL facilities – Mr Alexander Radostin

The Russian Fuel manufacturing company TVEL encompasses all the nuclear fuel cycle facilities in Russia, excluding mining. It fabricates nuclear FAs for more than 70 reactors in 30 countries from two facilities: MSZ (Elektrostal) and NCCP (Novosibirsk). The MSZ facility underwent upgrades in 2019, and the NCCP facility has recently completed its modernization in 2021, including the installation of a new manufacturing line for the TVSK fuel targeting the western PWR market. During these changes, process modifications were made to minimize the number of operations and automate as much as possible to limit human factors on production. Notable instances of this modernization include:

- The implementation of an automatic pellet visual inspection system;
- Automatic pellet loading in transport containers;
- Multidesign fuel rods manufacturing lines, increasing operational flexibility;
- Automatic skeleton welding.

These improvements have enabled TVEL to expand its manufacturing capacity and operational flexibility, enabling the company to effectively respond to the growing demand in WWER market, expand into the PWR market, and prepare for the utilization of future fuel types, including U–Pu fuels, ATF, and fuels with enrichments higher than 5% to support longer fuel cycles.

ID#16 China (CNNFC), Improvement of homemade coater – Mr Fan Diyang

CNNFC has developed and optimized a magnetron sputtering furnace to coat ZrB_2 on UO_2 pellets for the manufacturing of AP1000 type fuel. This sputter coater features 6 ZrB_2 targets or cathodes enclosed in a circular horizontal drum vacuum furnace. Several improvements have been implemented on this equipment including:

- New drum spindle design;
- Modification of the water pipes to cool down cathodes;
- Programmable logic controller (PLC)/software changes to eliminate certain bugs.

Following all these enhancements, the frequency of hardware and system failures has substantially dropped, and they are currently almost non-existent. This example demonstrates the CNNFC Baotou facility's capability to modify process equipment for the purpose of enhancing production stability and performance.

ID#17 China (CNNFC), Applications of machine vision technology in CANDU nuclear fuel production line – Mr Feng Huo

CNNFC has developed a new automatic visual inspection system to control the tube groove before tube welding in its CANDU manufacturing line. The welding of the endplug to seal the cladding tubes is a critical process for manufacturing quality fuel bundles, necessitating careful control of welding preparation to form a specific groove on the edge and prevent any welding issues. This new equipment combines several off-the-shelf vision control systems, such as the Keyence TM-3000 high speed camera with different types of lighting that are arranged in series. The coaxial light provides a clear image for dimensional measurement, while the scattered light provides information on the edge pollution. The primary advantage of such a control system is that it has high accuracy (with a success rate of over 90%), enhanced traceability, and is non-contact nature, which limits the risk of potential contamination before welding.

After the inspection of tube edges, a robotic arm was implemented to sort the tubes automatically based on their future position in the fuel bundle and transfer them to corresponding trays. This robotic arm is paired with a Keyence CV-X camera system to analyse the tubes in the box and communicate with the robot to sort it by bundle position. Moreover, the robot offers the additional benefit of using negative pressure suction to delicately manipulate the tubes, preventing any damage to them during the process. The increase in automation helps reduce costs by reducing the number of operators by around 30%. It is a trend that will be continued in China to move towards more intelligent manufacturing with the implementation of Industry 4.0 features.

ID#18 Spain (ENUSA), New technologies and innovations in quality inspection equipment at ENUSA's factory – Mr David Verdejo Garrido

In the same way, the ENUSA fuel manufacturing facility has incorporated innovative technologies to develop several automated inspection systems and is exploring the implementation of augmented reality to improve and streamline routine maintenance operations on the production lines. In the first presentation, ENUSA presented the development of two automated inspection systems to enhance the quality of its products: Automatic Pellet Inspection (API) for automatic inspection of pellets and 3DMPRO for the fuel rod surface inspection.

The pellet inspection equipment has been developed since the early 2000s with new generations of the system implemented since 2016 and more recently in the Chinese Jianzhong Nuclear Fuel Co. at Yibin manufacturing facility in 2019. The system utilizes high speed cameras to capture the image of the lateral surface of the spinning pellets on rollers, with the software automatically analysing the images for all types of indications or defects. A more recent innovation is the use of AI with convolutional neural networks to significantly improve software performance and accuracy. The key aspect of this development lies in training the model with an extensive database of images obtained by combining the direct images from both ENUSA and Yibin manufacturing facility, as well as utilizing data augmentation, which creates artificial images by modifying features of existing images. Through this approach, the system has achieved a commendable success rate of 90%, thus demonstrating the capacity and advantage of using neural networks to enhance quality inspection tools.

ENUSA developed and patented the 3DMPRO for inspecting the fuel rod surface using 4 laser profilometers with a high capture frequency enabling the inspection of the complete rod surface at speeds of up to 200 mm/s. This system can detect defects and provide a 3D computational view of them. A complementary manual system has also been developed to further characterize the defects detected online by the automatic system. Although this system could potentially be used to inspect other components, it would necessitate customized development for adaptation.

ENUSA has initiated the use of augmented reality for a particular regular maintenance operation, where operators wear glasses, and special software helps the operator detect and recognize specific parts while providing interactive guidance and especially traceability of the operations performed. Although this technology is still in its early stages of development, further testing of various applications may lead to its widespread adoption in the nuclear fuel fabrication industry.

ID#19 Spain (ENUSA), Passive scanner for uranium and gadolinium fuel rods inspection – Mr Javier Lafuente Sevillano

ENUSA also presented its double line passive scanner equipped with 126 gamma Bismuth Germanate detectors divided into 4 modules. This new scanner has replaced its previous combination of an active gamma scanner for UO_2 rods and a passive scanner for UO_2–Gd_2O_3 fuel rods. The main advantage of replacing an active scanner with a passive one is the removal of the ^{252}Cf, resulting in a significant reduction in both operating costs and operator dose, while also improving detection capacity and equipment stability. As part of this comprehensive equipment upgrade, the densitometer and eddy current inspection systems were also replaced with newer, more accurate and stable systems. Therefore, the replacement of the old equipment with the new inspection system adds significant value to ENUSA's fuel rod quality inspection process.

ID#20 China (CNNFC), Novel compact high speed passive gamma ray assay equipment for Uranium enrichment measurement of fuel rods – Mr Yongli Zhu

Similarly to ENUSA, CNNF has also developed a passive fuel rod gamma scanner with 128 silicon photomultiplier detectors divided into 4 modules, intended to ultimately replace its active scanner. This change offers the same benefits as described in the previous presentation (Section 2.4.5) in terms of cost reduction and operational flexibility. Moreover, the passive scanner has provided increased sovereignty due to the supply risk associated with the supply of ^{252}Cf, which is only produced in a few countries. However, the main downside of a passive scanner was its low throughput speed, which was compensated by a high number of detectors to enhance the signal, enabling the system to reach 6 m/min in this case. To maintain a high inspection accuracy, a constant speed has to be maintained as the fuel rod passes through all detectors, and the time delay of each detector has to be accounted for in the measurement. The presentation provided a detailed description of the calibration process using standard rods containing clearly identified pellets of different enrichments. It also discussed potential issues related to age correction due to the presence of ^{234}Pa. This new equipment has demonstrated its stability over the two years of operation and is expected to gradually replace the current active scanners.

2.4.2. Observations: Insights and Future Prospectives

The presentations show several trends for fuel manufacturing facilities:

(1) Increased level of automation in the production lines, as seen especially in the TVEL, CNNFC, and ENUSA presentations. The main benefits of automation are:
- Improved performance and efficiency;
- Increased operational flexibility;
- Reduced operational dose;
- Reduced human effects;
- Increased quality traceability.

(2) The development of new QC and inspection systems, which are often automated. The goal is to better evaluate the quality of the final product, but it is also used in intermediate manufacturing steps to enhance process control and improve the following fabrication steps:
- In this context, visual inspections are transitioning from human operations to automatic camera type systems;
- In some cases, the integration of AI with neural network image analyses is starting to be implemented.

(3) Regarding fuel rod gamma scan inspection, there is a global trend to move away from active gamma scanners toward passive ones. This was helped by the increase in computing speeds which now allows the simultaneous analysis of hundreds of detectors, compensating for the low detection speed of passive scanners to respond to the demanding manufacturing throughput needs. The main benefits of passive scanners, which do not require ^{252}Cf source are:
- Reduced operational cost;
- Sovereignty: no potential ^{252}Cf supply issues;
- Reduced operational dose;

- Increased production flexibility: can control both UO_2 and Gd_2O_3–UO_2 rods.

(4) Manufacturing equipment is continuously being upgraded to improve efficiency and performance.

The presentations in this session demonstrated the continuous improvement ongoing in all nuclear fuel manufacturing facilities aimed at fabricating new designs, to improve quality and safety, as well as general equipment performance and efficiency ultimately reducing fabrication costs. The incorporation of new technologies plays a significant role in plant upgrades, where automation is implemented whenever possible, thus limiting the impact of human factors on production but also reducing worker radiological dose. The first applications of neural network AI and augmented reality are being tested with promising results, suggesting a shift from usual conservatism in the nuclear industry towards the incorporation of recent less-proven technology. These emerging technologies present both significant opportunities and challenges as nuclear fuel manufacturing facilities evolve and prepare for the future needs of nuclear power worldwide.

2.5. TECHNICAL SESSION V: EXPERIENCE AND PROSPECTIVES FOR FUEL FABRICATION

Prepared by Mr Yeldos Kultayev (International Atomic Energy Agency)

This technical session was originally called "Fabrication technologies for new fuel development". It was followed by a discussion of prospects, challenges encountered, and areas that require international collaboration. The presentations highlighted the current state of affairs in fuel development, as well as prospects in some of the Member States. The common themes of the presentations included the history of fuel capacity development, current capabilities and their adaptation, descriptions of fuel types being developed and areas of research, R&D topics currently being explored, and challenges in ensuring fuel performance. In the presentation by Souza (ID#25), an analysis was performed to assess the adaptability of current fuel manufacturing capabilities to SMR fuels. The presentation by Alvarez (ID#21) provided information about current research activities in Argentina, such as improvements in the mechanical strength of fuel pellets, the replacement of Be brazing, enhancements in fuel resilience during power ramps, and the conversion of Atucha 2 NPP to slightly enriched uranium (SEU) fuel. The presentation by Yang (ID#22) discussed the challenges of deploying new generation fuels, particularly those associated with experimental data acquisition and suggested the use of unit tests with micro-machined samples and the adoption of a computational platform capable of accurately predicting fuel performance. The presentation by Rahmadi (ID#23) described the fuel development programme in Indonesia, where PWR, high-temperature gas reactor (HTGR), and heavy water reactor (HWR) fuel is being studied, along with various UMo compositions for research reactor fuel. The presentation by Marin (ID#24) shed light on MOX fuel utilization in France and upcoming new short term and long term designs under development. In addition, Arakelyan (ID#26) presented a project focused on justifying the use of 'Slim Rod' FAs at the Armenian NPP, while Almahmoud (#ID27) provided an overview of Jordan's nuclear programme.

2.5.1. Summary of presentations

ID#21 Argentina (CNEA), Argentine Experience – Mr Luis Alvarez

Mr Alvarez drew attention to the current nuclear infrastructure in Argentina, noting that the country's three operational NPPs exclusively employ PHWRs, with FAs manufactured domestically. This domestic production has enabled iterative improvements in design over time, with a strong emphasis on enhancing safety and economic viability. By leveraging this experience, Argentina developed local fuel engineering and manufacturing capabilities through initiatives like the CAREM Project. The author described the nuclear fuel industry in Argentina, which is characterized by three primary organizations (CNEA, Conuar-FAE and DIOXITEK, NASA), each with distinct responsibilities. The presentation outlined the primary areas of focus in Argentina's nuclear fuel development, which include:

- Material Sciences and Metallurgy, including the irradiation effects during the nuclear service;
- Product engineering, which involves simulating fuel behaviour during irradiation and spent fuel storage using computer codes;
- Manufacturing process development;
- Non-Destructive Testing and other QC techniques;
- QA;
- Post-Irradiation Examination (Pool Side Inspection and Hot Cell Techniques).

In conclusion, the author highlighted significant advancements in fuel technology development currently underway in Argentina, including improvements in the mechanical strength of fuel pellets, the replacement of beryllium brazing, enhancements in fuel resilience during power ramps, and the conversion of Atucha 2 NPP to SEU fuel. The author also noted that these fuel technology development activities have led to spin-off knowledge and services.

ID#22 Republic of Korea (KAERI), Nuclear fuel R&D for commercial reactors in KAERI – Mr Jae Ho Yang

Mr Yang delivered a comprehensive presentation on the historical trajectory of nuclear fuel capabilities development in Korea, ongoing R&D efforts at KAERI, and the challenges encountered in the deployment of advanced LWR fuel technology. The presentation suggested that KAERI is actively involved in a wide range of fuel development projects, covering both research reactors and power reactors. KAERI has been specifically engaged in the development of U_3Si_2 fuel for international partners, as well as plate-type U–Mo fuel for a new research reactor currently being constructed in Korea.

Within the domain of LWR fuel R&D, KAERI has divided its efforts into three main areas: improvement of existing technology, employment of innovative concepts, and development of ATF. Mr Yang also addressed the challenges associated with implementing advanced LWR fuel technologies, such as establishing fabrication capabilities and conducting long term irradiation tests in reactor-relevant environments to obtain experimental data. This undertaking requires significant investments in infrastructure and international collaboration to alleviate the financial burdens involved. As an alternative strategy, Mr Yang suggested using unit tests with micro-machined samples and adopting a computational platform capable of accurately predicting fuel performance.

ID#23 Indonesia (BRIN), Current status nuclear fuel programme in Indonesia – Mr Gagad Rahmadi

Mr Rahmadi, representing the National Research and Innovation Agency of Indonesia (BRIN), delivered a comprehensive presentation on the current state of the Nuclear Fuel Programme in Indonesia. The R&D efforts in Indonesia regarding fuel have encompassed both commercial NPP fuel and research reactor fuel, despite the absence of NPPs in the country. The foundation of nuclear fuel development in Indonesia began with a focus on HWR fuel, alongside the advancement of the Cirene HWR. Indonesia hosts three research reactors: Kartini, TRIGA MARK, and RSG G.A. Siwabessy, with the capability to produce nuclear fuel for the RSG G.A. Siwabessy research reactor.

Regarding the advancement of commercial NPP fuel, research efforts have progressed in line with the technology of interest. Currently, BRIN is engaged in PWR and HTGR fuel programmes. The PWR fuel programme includes research activities in various areas, such as UO_2 pellet fabrication and fuel pin fabrication. The research on UO_2 pellet fabrication has involved tests using powders with different enrichment levels to study properties such as O/U ratio, bulk density, tap density, powder size distribution, and moisture content.

The HTGR fuel programme has involved research aimed at assessing the fabrication process methodology, evaluating coating characteristics, and conducting heat tests on TRISO (Tri-structural Isotropic) particles.

In the context of activities related to the advancement of research reactor fuel in Indonesia, considerable attention has been given to the UMo alloy material within ongoing R&D efforts. Experimental trials that focused on the fabrication of mini-sized plate type dispersion fuel, containing UMo/Al, UMo/Al–Si, UMo–xM/Al, and UMo–xM/Al–Si compositions, have been conducted to determine the most suitable process parameters. This presentation contributes to a broader understanding of Indonesia's nuclear fuel developments.

ID#24 France (Orano), MOX fuel experience – Mr Jean-Michel Marin

In his presentation, Mr Marin provided an overview of the experience and prospects of MOX fuel within Orano and France. Beginning with a brief examination of global and domestic utilization of MOX, he outlined key milestones in Orano's MOX experience. One of the key milestones was the establishment of Orano's Melox Plant in 1990, dedicated to MOX FA production in France. The presentation introduced the French NPP fleet, which is technically equipped to utilize MOX FAs and mentioned that MOX utilization was practiced internationally. A comparison between the designs of MOX and UO_2 FA was provided, highlighting their similarities. The author also explained the fundamental distinctions between MOX and UO_2 FAs.

A comprehensive roadmap for MOX fuel development within Orano was presented, projecting MOX fuel deployment in 1300 MWe reactors by the 2020s, with a medium term objective of multirecycling MOX in PWRs by around 2040, subject to industrial feasibility. Furthermore, the presentation highlighted the potential benefits of fuel cycle closure through the utilization of Generation IV (GEN IV) fast reactors, including a reduction in waste volume and toxicity, as well as uranium savings. In alignment with these objectives, conceptual designs for MOX CORAIL and MIX FAs were presented as midterm and long term deployment targets, respectively.

ID#25 Brazil (INB), SMR fuel development and fabrication – INB Overview – Ms Patricia Souza

Ms Souza gave a presentation on SMR fuel development and fabrication, introducing the role of nuclear energy in Brazil's future and shedding light on the associated challenges in planning and investing in nuclear fuel production. A significant emphasis was placed on evaluating the current capabilities and importance of INB.

Additionally, the presentation examined the potential utilization of SMR concepts. Notably, reference was made to a study conducted by the Idaho National Laboratory, which explored the suitability of power reactor designs for the Brazilian market. The research conducted within INB aimed at evaluating light water SMR designs potentially manufacturable at INB facilities. The presentation of study outcomes included the identification of three selected PWR type SMR designs and a comparative analysis of their nuclear fuel technical parameters in relation to the FAs currently produced at INB.

The comparison revealed similarities in enrichment levels and technology platforms between the selected SMR designs and the FAs currently produced at INB. The evaluation indicated that only minimal adjustments would be required for powder and pellet fabrication processes, with slight impacts anticipated on skeleton welding, fuel rod assembly, and storage procedures, alongside significant implications for FA shipment and storage logistics. Souza concluded that Brazil's FA production facility has the potential to meet the future need for SMR FA production in Brazil and Latin America.

ID#26 Armenia (Haykakan Atomayin Electrakayan), Implementation of a new type of nuclear fuel at the ANPP – Mr Arakel Arakelyan

Mr Arakelyan, representing the Armenian NPP, delivered an overview of the Armenian NPP and recent upgrades in FAs to be utilized. Arakelyan presented a recent project dedicated to justifying the use of 'Slim Rod' FAs in Armenian NPP with the WWER-440 design. This Slim Rod FA was developed based on the design of the second generation FA. The detailed technical characteristics of the FA currently in operation were provided. Instead of the second generation fuel rods, which have an outer cladding diameter of 9.1 mm, the new FA is to use fuel rods with a reduced outer diameter of 8.9 mm. The dimensions of the fuel pellet remain the same at 7.6 mm, and the reduction in the outer diameter of the fuel rod results from a decrease in the thickness of the fuel rod cladding. Mr Arakelyan listed several advantages of implementing the Slim Rod fuel, such as optimal water-to-fuel ratio that leads to increased power generation and a reduced mass of zirconium used in the structural materials of the fuel rods, while maintaining the same dimensions of the fuel pellets.

ID#27 Jordan (Energy and Minerals Regulatory Commission), Nuclear programme in Jordan – Ms Amani Almahmoud

In her presentation, Ms Almahmoud provided an overview of the existing nuclear projects, such as the Jordan Subcritical Assembly, the Jordan Research and Training Reactor, and the exploration of Jordanian ores, as well as projects under study, including the NPP project in Jordan. Ms Almahmoud described the main reasons why SMRs were proposed for construction in Jordan such as expected short construction times, simplicity of design, and other factors, as well as the challenges involved in licensing the SMRs.

2.5.2. Observations: Insights and Future Prospectives

The summaries of the presentations provide a comprehensive overview of the development, challenges, and prospects of nuclear fuel programmes across different countries.

Efforts to localize fuel production, optimize fuel fabrication processes, enhance fuel performance, and develop advanced fuel technologies are evident across nations. Furthermore, some presentations (ID#21 and ID#22) highlighted the importance of collaboration between government agencies, research institutions, and industry stakeholders in advancing nuclear fuel programmes.

In the future, we may see more cases of collaboration between countries in fuel development to overcome the challenges associated with deploying advanced nuclear fuel technologies, such as infrastructure investments, long term irradiation testing, and innovative methods to predict fuel performance through the adoption of computational platform and the use of unit tests with micro-machined samples.

The presentations also emphasized the importance of diversifying nuclear fuel cycle options, including the exploration of SMR technologies and alternative fuel materials such as MOX fuel and UMo (Uranium–Molybdenum) alloys.

Although not every country's experience is covered, these presentations highlight the key insights into the development of fuel capacity building:

- — Fuel capacity building is a step-by-step process, mostly triggered by NPP construction and drive for self-reliance;
- — The import of fuel technology may be required at the early stages of nuclear energy programme development;
- — Fuel technology development can generate spin-off knowledge that can benefit society at large and contribute to industrial transformation;
- — Once there is a capacity in nuclear fuel manufacturing, it can be adapted to meet manufacturing demands to different designs.

The presentations were followed by a discussion session. The discussions were fruitful, with participants from Member States shedding light on the situation in the field of nuclear fuel development, focused on the following:

- — Developments in the adoption of ATF;
- — Developments in the use of AM technology (3D printing);
- — Fuel fabrication capacity building and localization efforts;
- — Efforts to develop technology for SMR fuel;
- — Efforts to adapt current capacity to work with enrichment levels higher than 5%, reprocessed uranium, and SMR fuel.

The utilization of AM technology has seen a significant rise in popularity. Initially, this technology was commonly used for prototyping, for example for optimizing the design of debris filters. In 2020, Westinghouse installed a Thimble Plugging Device, a safety related component, into an operational NPP. By 2022, Westinghouse had completed the installation of its StrongHold 3D printed nuclear fuel debris filters in BWRs in Finland and Sweden [7]. Furthermore, in 2024, a PWR bottom nozzle produced using AM technology was installed as a part of 4 Lead Test Assemblies at Farley NPP in the USA. On the other hand, in 2022, Framatome announced the completion of the installation of the first 3D printed upper tie plate grids of its ATRIUM 11 design FA, which was installed in Forsmark Unit 3 in Sweden [8]. Recently, Westinghouse revealed the production of its 1000th component using AM technology [9], demonstrating the potential for scaling up this technology for serial production.

3. CONCLUSIONS

There are two main drivers for improving the fabrication technology of nuclear fuel for commercial NPPs. The first is the desire to maintain nuclear fuel reliability with good performance during NPP operation. Poor reliability can lead to fuel failure in the reactor, resulting in high costs for failed core operation, maintenance, and increased dose rates for plant workers. Even if fuel failures do not occur in the reactor, nuclear fuel with poor reliability can create uncompetitive operational conditions, imposing significant economic penalties on a NPP operation. Oversight of fuel fabrication with improved/advanced technologies is one of the key contributors to maintaining nuclear fuel reliability and good performance in commercial NPPs. The second driver is to support new fuel development for existing and innovative NPPs. Efforts to enhance the safety, performance, and economics of the NPP are closely tied to the introduction of new fuel concepts. The fabricability of these new fuel concepts, with or without modifications of existing equipment and facilities, needs to be proven early in the development process. An example of this includes resolving welding issues between different materials.

Through two Technical Meetings held in 2021 and 2023, the IAEA has provided a platform for exchanging information on several advancements in fuel fabrication technologies to support the reliability, economics, and safety of power reactor fuels. Experts who participated in these meetings discussed achievements and ongoing progress in resolving challenges in the following technical areas:

- Powder and reconversion;
- Sintering, pelletizing and advanced fuel pellet fabrication;
- Fuel rod and assembly components including fuel cladding;
- Advancements in fabrication facilities and inspection;
- Fabrication for new fuel developments.

Presentations at both Technical Meetings indicated:

- Powder and reconversion:
  - Enhanced blending process for powder fabrication enables to reduce of the allowable tolerance range of homogeneity required for blended powder and to increase the content of plutonium in the MOX fuel;
  - Analytical approach (e.g. modelling & simulation) to powder preparation helps to ensure conditions that allow fuel pellets to best comply with technical specifications;
  - Improved reconversion processes provide economic benefits through the recovery of scraped uranium (wet process case) or through the reduction of liquid waste produced (dry process case).
- Sintering, pelletizing, advanced fuel pellet fabrication:
  - Analytical approach (e.g. modelling& simulation) to the sintering process helps determine optimal conditions for the microstructural characteristics of the fuel pellets;
  - The use of additives or dopants to improve fuel behaviour in terms of fission product retention, enhance thermal conductivity or incorporate neutron absorbers.
- Cladding fuel rod and FA components:
  - Several approaches have been explored to replace Be in the brazing of appendages to the cladding due to health and environmental issues. One is to use an alternative

material that has been confirmed as ready for use in CANDU fuel bundles, while another involves attaching the appendages by resistance welding;
  - Developments to enhance the performance of fuel cladding include new Zr based and other alloys, as well as the use of coatings applied on Zr alloys.
- Advancements in fabrication facilities and inspection:
  - Streamlining the fuel fabrication process with automated inspection systems increases manufacturing capability and enhances the operational flexibility of the facility to accommodate various market demands.
- Experience and prospectives for fuel fabrication:
  - Self-reliance on nuclear fuel supply is pursued by several nuclear embarking countries as part of their plan to expand nuclear power, including the development of SMRs.

From the above, discussions held during the Technical Meetings, it is concluded that:

- Recent improvements in powder preparation technology and increased safety of fabrication plants facilitate the manufacture of advanced fuel cycle fuels that may contain multiple fissile materials or higher enrichments;
- Recent improvements in fuel rod and assembly manufacturing processes (e.g. streamlining processes with automated inspection systems) enable the production of reliable and robust fuels for operating reactors and increase the operational flexibility of fabrication facilities to accommodate market demands on various fuels with different designs;
- The rising demand for ATFs combined with LEU+/HALEU, and advanced fuel cycles (e.g. reprocessed uranium), driven by safety and economic considerations, will require reconsideration of current processes and practices. This may include the adoption of cutting-edge technologies such as AI, augmented reality and AM.

Recognizing the substantial investments expected in fuel fabrication technologies and facilities to advance the development of innovative reactors, the meeting participants emphasized the importance of platforms, such as workshops, to foster the exchange of information on fuel fabrication technologies..

EXTENDED ABSTRACTS AND FULL PAPERS SUBMITTED AND PRESENTED AT THE TECHNICAL MEETINGS

TECHNICAL SESSION I: POWDER AND RECONVERSION

ID#1

DEVELOPMENT OF AUH (AMMONIUM URANATE HYDRATE) WET RECONVERSION PROCESS FOR THE PRODUCTION OF NUCLEAR-GRADE UO_2 POWDER FROM URANYL NITRATE SOLUTION

B. LEE, S. YANG, D. KWAK, H. JO, Y. BAE
KEPCO Nuclear Fuel (KEPCO NF),
Daejeon, Republic of Korea

Abstract

This paper provides a brief description of the AUH process, which is a new wet reconversion process that KEPCO NF has developed. This AUH process has improved characteristics compared to the existing wet reconversion process. With simplified process steps, the AUH process produces large and free-flowing UO_2 powder, and the powder satisfies all DC pellet specifications including the sintered density and density change after the re-sintering test. Therefore, this new process can be utilized as a manufacturing process for nuclear-grade UO_2 powder.

1. INTRODUCTION

KEPCO NF operated the AUC process as the main UO_2 production process until 1997 and converted the main process to the DC process in 1998 because the AUC process emits a large amount of liquid waste. Additionally, KEPCO NF had operated the ADU process to recover uranium scrap for several years but failed recovery though the process worked normally because KEPCO NF did not have a uranium purification process at that time.

In the 2000s, KEPCO NF faced with a problem. Uranium scrap accumulated for 20 years, and the inventory exceeded 20 tonnes by uranium content (tU). And the annual generation was increasing according to the production increasing. However, KEPCO NF could not recycle scraps, because KEPCO NF did not have a uranium purification process nor a wet reconversion process. Thus, the project to develop both processes started in 2010.

As mentioned above, KEPCO NF has experienced 2 wet reconversion processes and learned many things from this experience: AUC process produces large (over 20 µm) and free-flowing UO_2 powder easy to handle, and this gives many advantages for operation. But AUC is slightly soluble in water, so the filtrate contains about 1000 ppm of uranium and the filtered cake cannot be washed with water. This is a big disadvantage. On this account, AUC process needs a large scale wastewater treatment facility to recover uranium in filtrate, and a methanol recovery system for cake washing. Thus, AUC process has complicated process steps including explosive liquid treatment and needs a large scale plant relative to its capacity. On the other hand, the ADU process produces UO_2 powder which shows better sintering performance than from AUC process. However, the powder is agglomerating fine particles and does not flow, so it gives us a big disadvantage for operation. And ADU process still has complicated process steps, and high uranium content in filtrate even though it is lower than AUC process.

Therefore, this is why KEPCO NF needed to develop a new wet reconversion process, namely "AUH process", together with a uranium purification process to recycle uranium scraps. The purpose of the project was to improve the characteristics of a wet reconversion process to avoid the disadvantages of existing processes. Especially, simplifying process steps, improving powder characteristics, and lowering U content in filtrate.

In this project, KEPCO NF developed this process and produced materials, and those have different features, based on the analysis results, compared to the existing wet processes, AUC or ADU. Thus, this process was denominated as AUH wet reconversion process after the name of the intermediate material, AUH, like AUC or ADU processes.

2. PROCESS DEVELOPMENT AND EXPERIMENTS

2.1 Development and design

To recover uranium from purified uranium nitrate solution produced from the Uranium purification process, the target process scheme was formulated as Crystallization (with ammonia) — Liquid–Solid (L/S) separation — Fluosolids reduction (with nitrogen and hydrogen).

The reactor type was determined as slab type vessel for the Crystallization and a fluidized bed reactor for the Fluosolids reduction. These reactors require lower fabrication costs than the other types like mixer-settler and rotary kiln because they have no mechanical driving parts and occupy less space. In addition, the supplied gas flow to the fluidized bed reactor was designed in the form of a square pulse to overcome complications such as plugging, cracking, and channelling.

Through various lab scale experiments, some process parameters to test were selected, and pilot scale experiments were performed with a variation of parameters.

2.2 Pilot scale experiments

For the experiments, a purified UNH solution of 205 g/L was prepared by the Uranium purification process. Using this solution, 4 batches of crystallization experiments were conducted. This step is similar to the typical ADU process, except for the feeding method of ammonia. A gaseous air-ammonia mixture was introduced through a bubbling nozzle located at the bottom of the reactor until the target pH was attained. For all batches, the initial concentration of uranium and reaction temperature were fixed. The target pH was 8 for the 2nd batch and 7.2 for the other. Air-to-ammonia ratio was 1.4 for the 2nd batch and 0.7 for the others.

Upon the completion of the reaction, the final slurry was filtered under vacuum and dried under hot air flow. For this L/S separation step, there was no particular development but a noticeable difference in the developed crystallization step from that of ADU process was found: the operation time. It took 20–30 minutes for filtration of 16 kgU of AUH slurry and it became dried AUH powder after 12 hours of drying at 150°C, while it takes generally over 2 hours for 15–20 kgU of ADU slurry and the filtered ADU cake cannot be dried by drying over 24 hours at the same temperature.

The obtained dried AUH powder was converted into UO_2 powder in the next step, the Fluosolids reduction. This step is a heat treatment consisting of 3 stages: calcination, reduction, and stabilization. Calcination was conducted under an inert atmosphere with heating. During reduction, nitrogen-hydrogen mixed gas was supplied under the scheduled reaction temperature profile. The final calcination temperature was about 600°C, and the reduction temperature was about 800 °C for 1st batch and 560 °C for the others. The powder was cooled under a nitrogen atmosphere after the reduction reaction was completed, and nitrogen-air mixed gas was used for stabilization. Furthermore, finally, 4 batches of UO_2 powder were obtained from the pilot scale experiments of the developed AUH process.

3. RESULTS AND DISCUSSION

3.1 Chemical composition of AUH

After the L/S separation step, the dried AUH powder was sampled and analysed with Energy-dispersive X ray spectroscopy for estimation of chemical composition, because it showed a different appearance and powder behaviour compared to those of AUC or ADU. From the result shown in Table 1 [1], the empirical formula of AUH was determined as $UO_3{\cdot}0.16NH_3{\cdot}5.33H_2O$. As shown in Table 2 [2], the chemical composition of AUH is different compared to that of ADU.

TABLE 1. ENERGY-DISPERSIVE X RAY SPECTROSCOPY ELEMENTAL ANALYSIS OF AUH

Element	Composition (wt%)			
	Point 508	Point 511	Point 512	Average
Oxygen	17.8	18.19	19.01	18.33
Carbon	0.65	1.43	0.67	0.92
Nitrogen	0.75	0.82	0.73	0.77
Uranium	80.8	79.56	79.59	79.98

TABLE 2. CHEMICAL COMPOSITION OF THE ADU

Composition type	Chemical composition	Uranium content (wt%)
I	$UO_3 \cdot 2H_2O$	73.9
II	$UO_3 \cdot 0.33NH_3 \cdot 1.67H_2O$	74.0
III	$UO_3 \cdot 0.50NH_3 \cdot 1.50H_2O$	74.0
IV	$UO_3 \cdot 0.67NH_3 \cdot 1.33H_2O$	74.1

3.2 Particle size distribution of AUH and converted UO_2

In the Crystallization step, AUH slurry was sampled and analysed with the particle size analyser. In Figure 1, the fluctuation of the mean particle size as a function of pH is presented [1]. Figure 1 illustrates that the average particle size falls within the range of 10–20 μm after the reaction is completed.

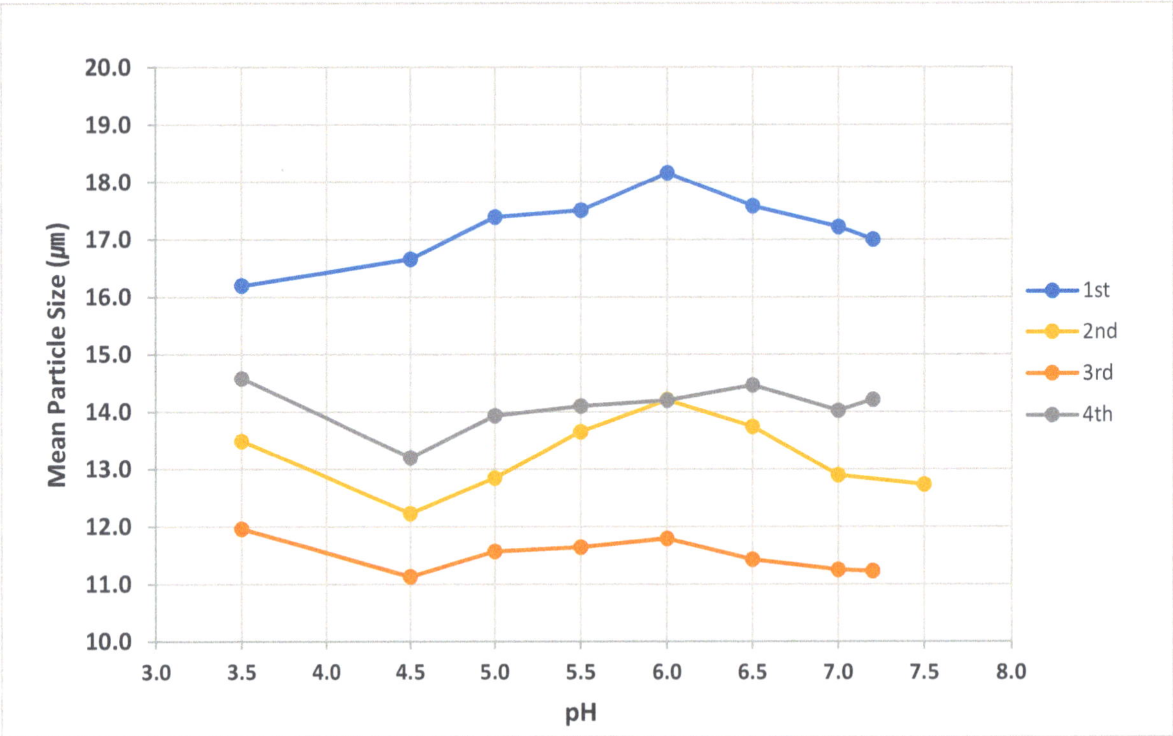

FIG. 1. Variation of mean particle size of AUH with pH.

Figure 2 shows particle size distributions after the crystallization reactions were completed and Figure 3 shows those of ex-AUH[1] UO_2 powders. For all batches, the mean values are over 10 μm and the minimums are over 1 μm for AUH. After the conversion of AUH to UO_2, the particle size slightly decreased but still showed similar distribution to that of AUH powder.

[1] The prefix "ex-" means "converted by the corresponding process" in this paper.

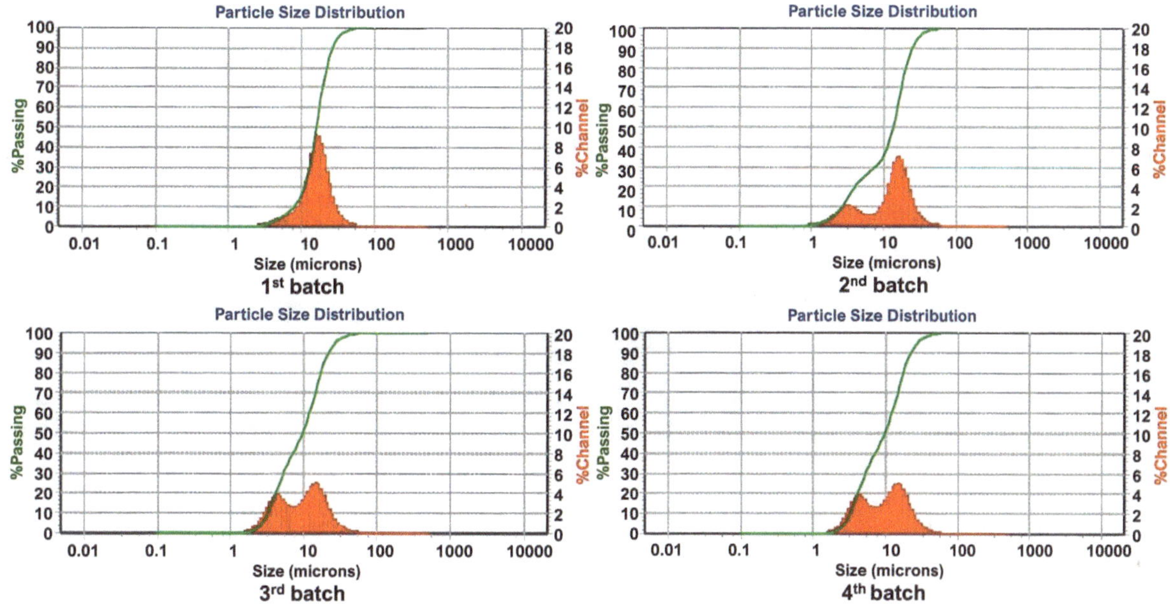

FIG. 2. Particle size distribution of AUH powder.

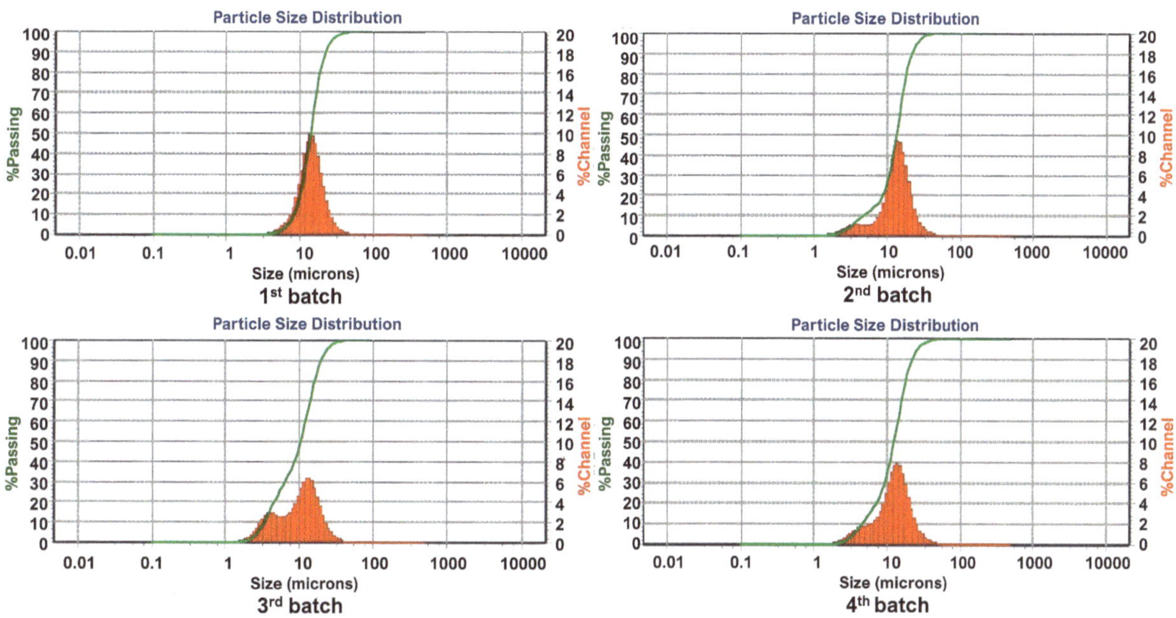

FIG. 3. Particle size distribution of ex-AUH UO_2 powder.

3.3 Ex-AUH UO_2 particle morphology and powder characteristics

Figure 4 shows the scanning electron microscopy (SEM) images of ex-AUH UO_2 powder. Particles have skein-like features on the surface and inside. The shape is different from those of ex-AUC and ex-ADU UO_2 powders.

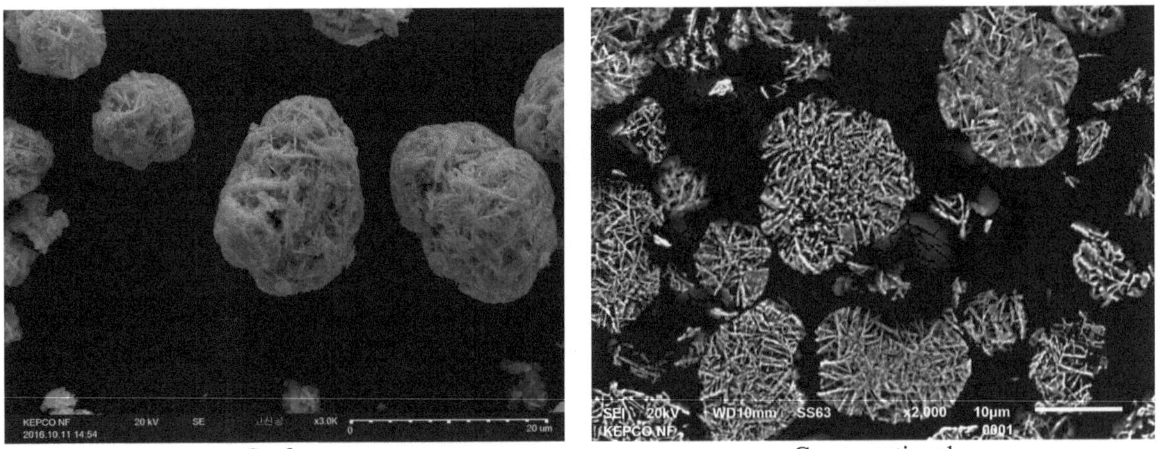

Surface Cross-sectional

FIG. 4. Particle shape of ex-AUH UO_2 powder.

Table 3 [1] provides a comparison of the properties of UO_2 powder from three different processes mentioned above. It is shown that ex-AUH UO_2 powder has a relatively high BET surface area. It seems that this is due to the skein-like structure.

TABLE 3. COMPARISON OF UO_2 POWDER PROPERTIES

Properties	Ex-AUH	Ex-AUC	Ex-ADU
Specific surface area (m^2/g)	6.93	5.00	3.00
Apparent density (g/cm^3)	1.50	2.30	1.41
O/U ratio	2.19	<2.15	<2.14

3.4 Sintering performance test

The result of the sintering performance test is shown in Table 4 [1]. The test was conducted as per the DC pellet specification, for there is no specification for ex-AUH UO_2 powder yet. The third (3rd) and fourth (4th) batches satisfied the specification and the result leads to the conclusion that ex-AUH UO_2 powder can be applied to nuclear fuel. Additionally, ex-AUH UO_2 powder shows good sintering properties though its O/U ratio is higher than those of ex-AUC and ex-ADU UO_2 by about 2%. Thus, it is expected that the wider O/U ratio range can be allowable for ex-AUH UO_2.

TABLE 4. EX-AUH UO_2 POWDER PERFORMANCE TEST

Batch No.	Sintered density (g/cm^3)	Rate of increase of re-sintered density (%)	Grain size (μm)
1	9.64	2.14	-
2	10.27	1.13	9.68
3	10.41	0.48	10.34
4	10.42	0.61	8.32
Specification	10.30 ~ 10.58	≤ 1.0% TD	≥ 5.00

A comparison of the AUH process with other processes is shown in Table 5 [1]. One of the great improvement points is the uranium content in AUH filtrate. In this pilot scale test, the filter unit used for the commercial ADU process was utilized as it was, and the filter cloths were the same. Uranium contents in AUH filtrate were 10 ppb for 3 batches out of 4 and the rest was 2.7 ppm while it used to be up to 130 ppm for the ADU process. For AUC process, it was about 800–1000 ppm. This result shows that AUH process can reduce the scale and steps of wastewater treatment process compared to the ADU or AUC process.

TABLE 5. COMPARISON WITH OTHER PROCESSES

	AUH	AUC	ADU	DC
Process steps	Moderate	Complicated	Complicated	Simple
Avg. particle size (μm)	13	20	2	2
BET (m^2/g)	7	5	6	3
Flowability	Good	Good	None	None
Powder pre-treatment	Not required	Not required	Required	Required
Sintered density (%TD)	95	95	96	96
U content in filtrate (ppm)	< 0.1	~1000	~130	-

AUH process is much simpler than AUC and ADU but slightly more complex than DC. AUH has a larger particle size, BET, and good flowability, therefore, AUH powder and ex-AUH UO_2 powder are easier to handle, and an additional powder pre-treatment process is not required, unlike ADU and DC. Sintering properties are similar to the others.

4. CONCLUSIONS

AUH process is a much simpler wet reconversion process in comparison with AUC or ADU process. The biggest difference is that AUH process does not need process steps to recover U from the filtrate because U content in the filtrate is very low.

UO_2 powder produced from AUH process is free-flowing, has large particle sizes and high BET, and has good sintering properties.

Accordingly, AUH process can be applied to the manufacturing of nuclear-grade UO_2 fuel.

REFERENCES

[1] LEE, B., et al., Ammonium uranate hydrate wet reconversion process for the production of nuclear-grade UO2 powder from uranyl nitrate hexahydrate solution, J. Nucl Eng Technol **55** (6) (2023) 2206-2214.

[2] WOOLFREY, J.L., The effect of pH on the properties of ammonium uranate precipitated with gaseous ammonia, Australian Atomic Energy Commission (AAEC), **E397** (1976) 4.

ID#2
MODELLING OF DRY CONVERSION PROCESS

C. NICOLLET[*], J. BISCHOFF[*], A. MARIE[*], F. VIRY[**], P. NAMY[**]
[*] Framatome,
Romans-sur-Isère Cedex

[**] SIMTEC,
Grenoble

France

Abstract

The conversion of UF_6 to UO_2 at the Romans-sur-Isère Framatome fuel manufacturing plant uses the DC process, which limits liquid contaminated wastes and increases productivity. This process is complex with chemical solid gas reactions involved, and it is difficult to always know all the reactions taking place in the reactor and calciner, therefore Framatome has launched a project to model it. This modelling has several benefits:

- Perform parametric studies without using the production equipment;
- Increase performance by optimizing the process;
- Teaching tool through theoretical description of the process and its key influencing parameters.

The modelling is a step-by-step approach, adding new reactions or functions at each step. First, the geometry of the conversion reactor was established with the general modelling of the gas flows before focusing on the chemical reactions occurring at the gas nozzle, then adding the temperature influence, and finally the UO_2F_2 particle precipitation. The model created is then validated with experimental data measured during the reaction. Parametric sensitivity calculations can be performed to evaluate the influence of different parameters on the process and therefore optimize the reactions. For example, the N_2 flow rate can be varied, and we can see the impact on the distance from the injector at which the UO_2F_2 particles are created. This helps determine the appropriate flow rate range to limit particle accumulation on the injector, which would block it, and make sure the particles are not formed on the reactor walls. The use of modelling thus enables better control of the process.

1. INTRODUCTION

Framatome has nuclear fuel manufacturing plants in three countries: France, Germany, and the United States of America. A transverse and global fuel manufacturing R&D was set up to continuously improve and optimize its fabrication processes with the aim to increase performance and deliver quality products to its customers. In all its manufacturing plants, to transform enriched UF_6 to UO_2 powder used for pelletizing Framatome uses DC processes, which limit liquid contaminated wastes and increase productivity. Two technologies are used: one with a fluidized bed for the US and German plants and a direct conversion process for the French plant. This article will focus on the French DC process, which involves complex chemical solid gas reactions.

Modelling the equipment and process is a part of Framatome's approach mastering all the parameters and all of the phenomena which impact or occur during the production of the powder. Depending on the level of accuracy that we need, this work can be time consuming. To optimize this computing time, a step-by-step approach was used.

Each step has to be validated experimentally and is then coupled with subsequent steps to have a model with the best accuracy we can. The different steps of the conversion reactor modelling were done in the following order:

- Build a geometrical model: Define all of the parts which compose the reactor, the physical characteristics of the equipment, and the size of the mesh that we need to obtain a good accuracy of the model;
- Aeraulic Model of the gas flow: Define all of the gas movements and the gas speed field inside the reactor;
- 2D-axisymmetric Gas Injector Thermochemical Model: Implement injector gas flows with the chemical reaction, thermal impact of the reaction, and thermal catalytic reaction feedback;
- Integrate the thermochemical model at the reactor scale;
- Model the UO_2F_2 particle distribution in the reactor: Define the movement of all of the particles depending on the size.

As described in the following chapter all this work is performed as a compromise between the accuracy of the model and the computing time.

2. CREATION OF A GEOMETRICAL MODEL

The goal is to simplify the geometrical model in order to limit the time of the calculation and perform fine meshing on the specific areas of interest. Depending on the difficulties of the equations that are used and the level of detail needed, the calculation time can increase dramatically. The method applied is firstly to adjust the size of the mesh locally where the interesting phenomena occur (chemical reactions, interactions between the different gas flows) and secondly to adapt the size of the mesh at the good level of accuracy of the model. Finally, it is a compromise between the result obtained and the computing time needed.

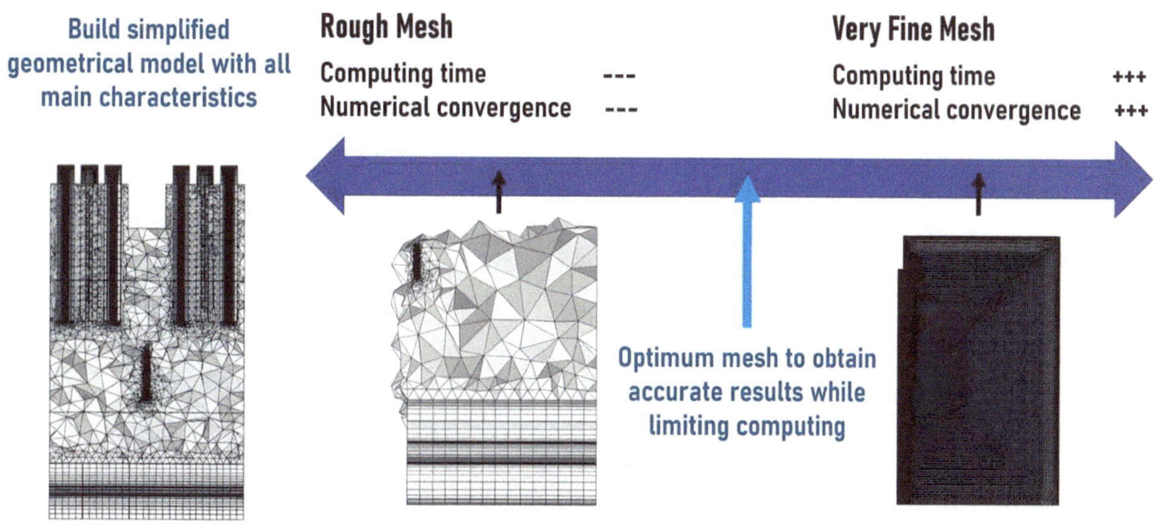

FIG. 1. Modelling of the geometry of the reactor.

As shown in Figure 1, several levels of mesh are tested step by step ranging from a rough mesh to a very fine mesh. For each level of mesh, the calculations are performed and compared until the result variations become negligible and the computing time is not too long. This becomes the reference mesh and model.

Figure 1 also shows that the mesh is very rough when there are not many variations of the physical characteristics of the medium (middle of the reactor) and very fine near the parts where the reaction occurs.

3. MODELLING OF THE GAS FLOW

The reactor contains two gas inlets (injector and gas coming from kiln/calciner) and one outlet through the filter candles. UF_6, H_2O, and N_2 arrive at the injector for the chemical reaction, an H_2/H_2O mixture used for the reduction of UO_2F_2 to UO_2 flows in the reactor from the pyrohydrolysis kiln through the Archimedes screw at the bottom (Figure 2). The chemical reaction forming UO_2F_2 leads to the production of HF gas that exits the reactor through the filter candles.

To model properly the gas flow, it is necessary to determine the flow regime (laminar or turbulent for example) in the two main parts of the equipment: the reactor and the filter candles. The calculation of the Reynolds number has been done with the following equation:

$$Re = \frac{\rho u L}{\mu} \tag{1}$$

where

- ρ is the volumic mass;
- u is the average of the speed of the gas;
- μ is the dynamics viscosity of the gas;
- L is the characteristic length of the equipment, here where the gas pass throughput.

For both the reactor and filter candles, the Reynolds number is around 8000 confirming a turbulent flow regime. This kind of regime is very difficult to model simply with a Direct Navier-Stokes approach, in which a too fine mesh could be required. A simpler model therefore has to be applied for the calculation and in this case, the RANS k- ε [1] model was chosen. It is a "Reynolds Averaged Navier-Stokes" model.

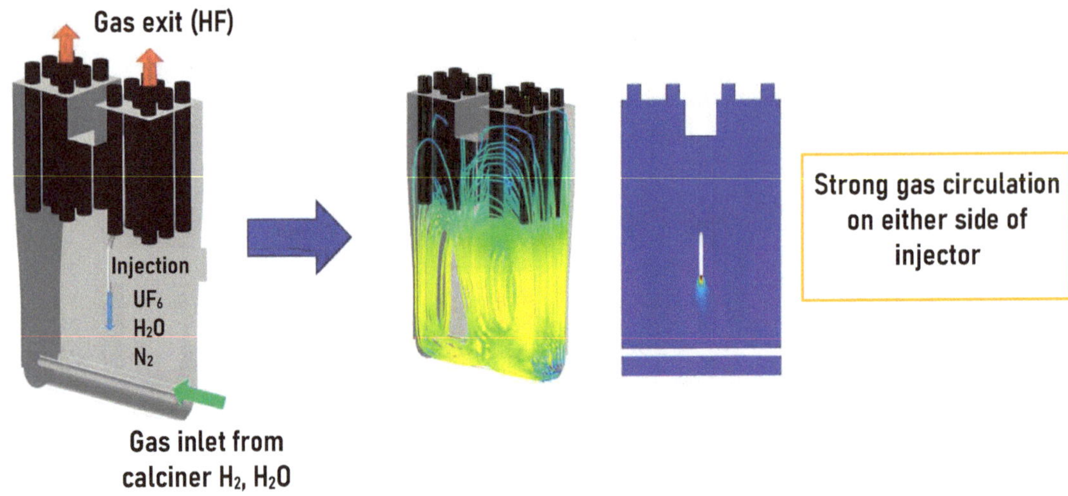

FIG. 2. Speed field of the gas inside the reactor.

As a first approximation, the calculation has been performed considering that all gases are the same homogeneous non-compressible gas, with the same viscosity, and same density. This hypothesis has been done to validate the overall speed field of gases in the reactor taking into account all inlets and outlets, with gas inlets from the pyro-hydrolysis kiln and injector, and outlets through the candle filters. The model established shows some vortex on either side of the injector which can cause some attrition between the surface of the reactor and the UO_2F_2 crystals generated at the injector nozzle, probably leading to some powder accumulation on the reactor surfaces. Such powder accumulations were observed physically on the equipment during maintenance, which helped confirm and validate the numerical model.

4. MODELLING OF THE REACTION AT THE INJECTOR NOZZLE

At this step, the kinetic parameters of the chemical reaction have to be defined in order to model the reaction taking place at the exit of the nozzle. The pyro-hydrolysis reaction between UF_6 and H_2O is very fast and has therefore been considered instantaneous in many studies [3]. For modelling, and especially to see where the reaction occurs downstream from the nozzle, it is important to use good kinetic parameters, but they are very difficult to define accurately so rough estimations followed by sensitivity studies were performed. The methodology is to calculate the effect of different parameters on the reaction flame and validate the results by comparing them to the ones observed experimentally. Two main results were used for this calibration: the temperature of the reactor and the localization of the crystals when the nitrogen flowrate is increased at the nozzle.

The transport-reaction process of chemicals is modelled using a convection-diffusion-reaction equation.

$$\nabla \cdot (\boldsymbol{u} c_i - D_T \nabla c_i) = \nu_i \cdot k \cdot c_{UF_6} \cdot c^2{}_{H_2O} \tag{2}$$

Term of convection | Diffusion of the reactants | Speed of the reaction

The first part of the equation describes convection and diffusion phenomena and the second part the evolution of the chemical reactants. Diffusion is mainly due to turbulence and is modelled with a Fick law. Here, the stoichiometric composition was used for the concentration of steam and UF_6. k is the reaction kinetic coefficient and follows an Arrhenius law, and as stated previously, it is an unknown parameter difficult to quantify precisely.

To determine a rough estimate of this kinetic coefficient, a sensitivity calculation study was performed to evaluate the influence of k on the integrated amount of UO_2F_2 produced and on the distance from the nozzle (Figure 3) where the total reaction forming UO_2F_2 takes place as shown in Figure 4. It is important this distance is not on the Archimedes' screw nor on the nozzle itself to prevent any powder accumulation on these mechanical parts.

$$UF_6 + 2H_2O \rightarrow 4HF + UO_2F_2$$

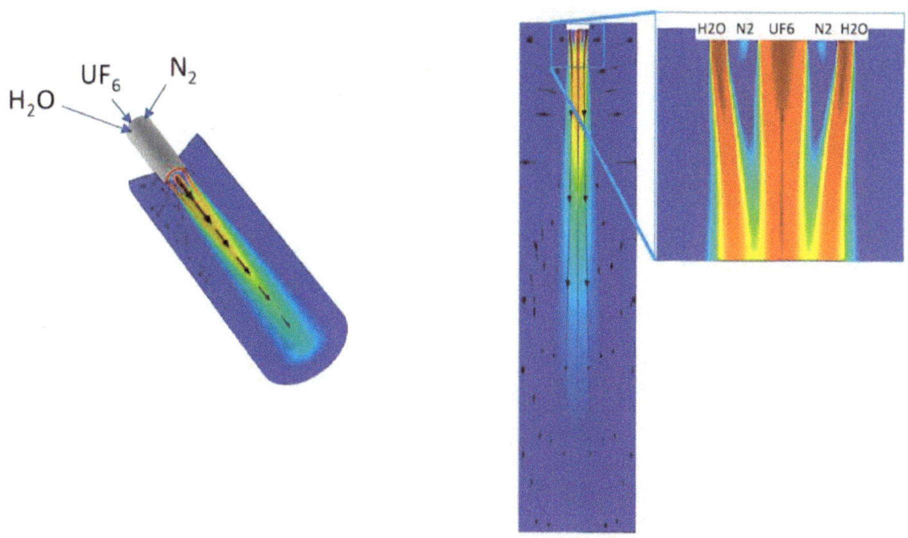

FIG. 3. Gas flow near the nozzle.

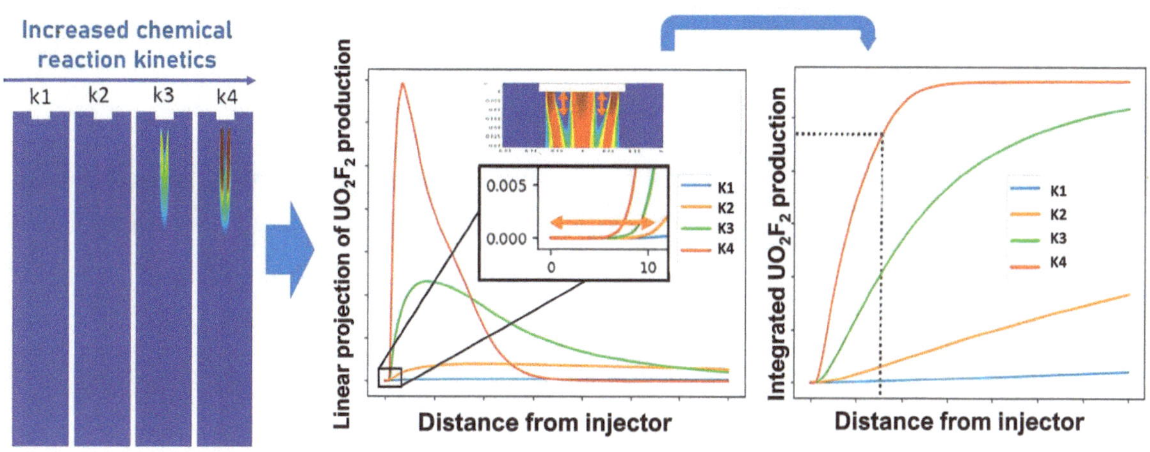

FIG. 4. Evaluation of the distance from the nozzle to have total reaction.

In Figure 4, the comparison of different k coefficients shows that the $k4$ coefficient is the good order of magnitude for the kinetic coefficient because all the others describe a crystallization on the surface of the Archimedes' screw of the reactor, which is not what is observed in case of normal flowrate of nitrogen. k is dependent on the temperature, and the chemical reaction is exothermic, leading to a catalytic effect due to the thermal-chemical coupling. The chemical reaction will engender a local increase in temperature, which will itself increase the reaction kinetics with an increase in UO_2F_2 production thus looping back to the temperature. Consequently, once the chemical reaction is modelled, it has to be coupled with the thermal one to obtain more accurate results. Figure 5 shows the impact of the coupling on the temperature of the reaction flame, which increases locally due to the coupling after the nozzle where the majority of the chemical reaction takes place. The coupling has no effect further away from the nozzle where no chemical reaction takes place.

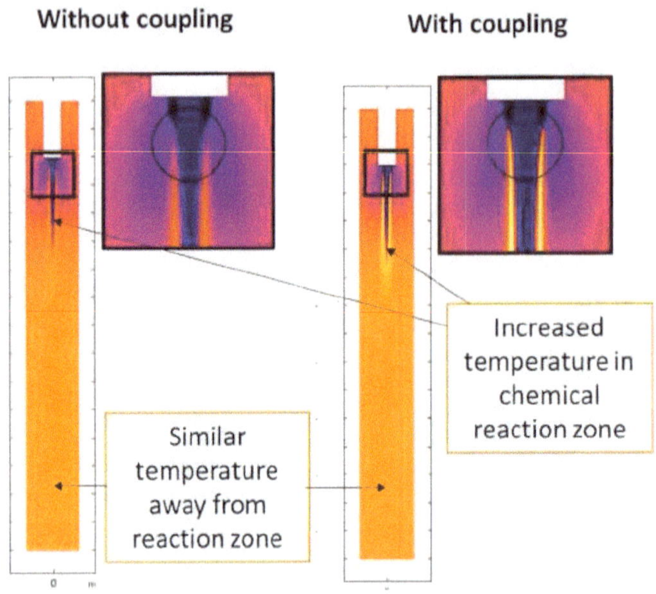

FIG. 5. Comparison of the model of the flame with and without coupling the temperature.

By increasing the reaction kinetics, the coupling not only influences the temperature but also the distance from the injector at which the total integrated production of UO_2F_2 occurs. Figure 6 shows that this distance is almost divided by 2 when the coupling is implemented compared to without the thermal-chemical coupling. Once this coupling has been successfully implemented, another sensitivity study on the k coefficient can be performed to obtain more accurate values.

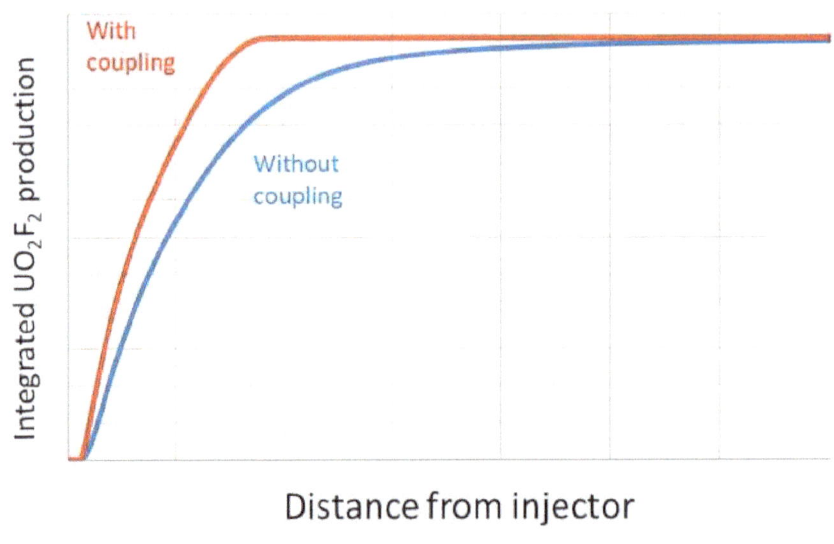

FIG. 6. Integrated UO_2F_2 produced vs the distance from the injector with and without coupling the temperature.

5. COMBINATION OF REACTOR GAS FLOW AND INJECTOR NOZZLE MODELS

Once the 3D reactor gas flow and the 2D-axisymmetric thermochemical models have been successfully put in place, the next step is to combine both to obtain an overall process model, which can be adjusted and validated with empirical values. There are many thermocouples on the reactor in order to survey operation and the temperature values measured can be compared to the values calculated by the overall model in order to validate all of the hypotheses taken while building the models. The nozzle and chemical reaction from it are considered as a heat source, which is dissipated by the gas flow inside the reactor.

Figure 7 shows in a cut plane that the temperature around the flame is lower once combined with the actual reactor flow than without.

The accuracy of the combined model has been validated with the comparison of the real temperature measurement inside the reactor (four thermocouples), near the nozzle (three thermocouples), and on the surface of the reactor (five thermocouples). The differences between the measured and the calculated values are on the order of 10 to 20°C, taking into account the isolation cover around the reactor and the thermal exchange with the outside.

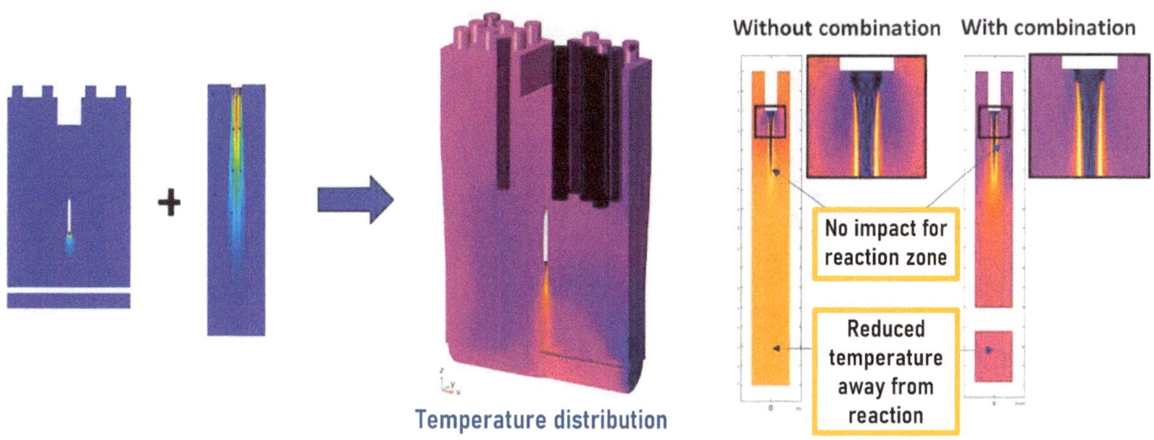

FIG. 7. Distribution of the temperature into the reactor after coupling the thermochemical model with the reactor gas flow.

6. MODELLING OF UO_2F_2 PARTICLE DISTRIBUTION

The final step of the process modelling is to simulate the UO_2F_2 particle distribution in the reactor. For this modelling, it is important to consider many different characteristics of the particles produced. To model the particle evolution, Newton's law for motion is applied to each particle. This law depends on the particle geometry and density characteristics as well as gas and gas-particle interaction parameters summarized in the following drag force equation [2]:

$$\boldsymbol{F}_d = m \cdot \frac{3\mu_{\text{gas}} Re_r}{4\rho\varnothing^2} \cdot C_D \cdot (\boldsymbol{u}_{\text{gas}} - \boldsymbol{v}) \tag{3}$$

where

F_d	is the drag force;
C_D	is the drag coefficient;
V	is the speed of the particle;
ρ	*is* the volumetric mass of the particles;
$\varnothing$	is the particle diameter;
Re_r	is Reynolds number;
μ_{gas}	is dynamic viscosity of the gas;
u_{gas}	is the speed of the gas and v the speed of the particle.

The drag force coefficient is calculated by the law of Schiller-Naumann [4]. It has been shown during the test that the geometrical morphology of the particles has a minor influence over the other parameters since only the diameter is taken into account. As shown in Figure 8, many different geometrical forms give the same diameter. Consequently, as a first approximation, all the crystals and agglomerates of UO_2F_2 formed in the reactor were defined as spherical particles.

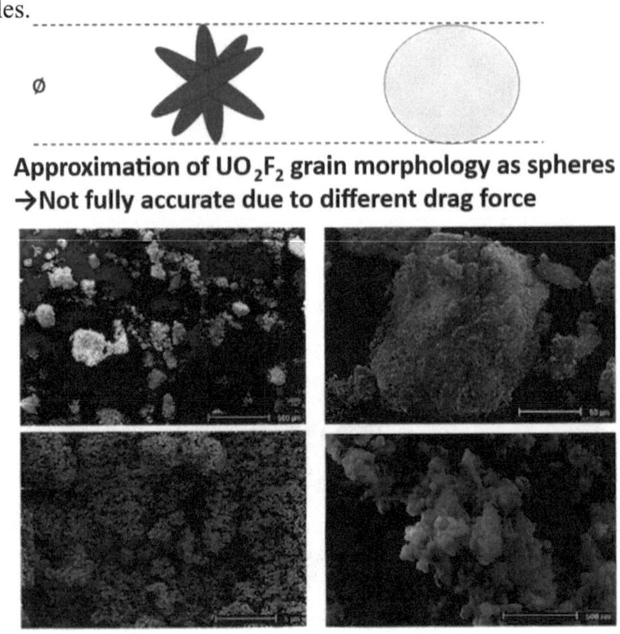

FIG. 8. Example of different UO_2F_2 particles with SEM.

From then, different UO_2F_2 particle densities and sizes were tested in sensitivity studies since there is a range of both in the reactor. The objective is to evaluate the distribution of each of these different types of particles in density and size inside the reactor and compare it with experimental values. For example, it was shown that the smallest particles have a tendency to accumulate on the candle filters, while the larger ones fall down directly on the Archimedes' screw. With this distribution of fine particles on the candle filters, some optimizations of the blow back sequence can be performed to prevent clogging of these candle filters, thus improving the overall performance of the equipment.

7. CONCLUSIONS

The DC process reactor is an equipment that can reach high throughputs to convert UF_6 into ultimately UO_2 but that is complex to operate. Many phenomena occur during production. To improve reactor performance in both powder quality and production throughput, a model was built to simulate the overall process with a combination of aeraulic and thermochemical models. A multi-step approach was taken to build the overall process model sequentially, with first a general 3D gas flow model, which was then combined with a 2D thermochemical model to find the best compromise between model accuracy and computing time. Each of these models can also be used separately to perform sensitivity studies focusing on specific parameters without using a big time-consuming larger model.

In the end, the overall process numerical model leads to a better understanding of the different phenomena taking place during production and enables parametric studies to improve the equipment performance either by reducing maintenance issues or by increasing throughput. Such a model is also a great tool for training production/maintenance operators and engineers to help visualize the phenomena taking place within equipment.

This R&D development is part of Framatome's dedication to continuous improvement to increase performance and deliver quality products to its customers.

REFERENCES

[1] JONES, W. P., LAUNDER, B. E., The prediction of laminarization with a two-equation model of turbulence, Int. J. Heat Mass Transf., **15** 2 (1972) 301-314.

[2] COMSOL.COM, COMSOL 5.6. Particle Tracing Module Users Guide, 268.

[3] CHOTARD, A., et al., Fabrication, characteristics, and in pile performance of UO2 pellets prepared from dry route powder, Key Engineering Materials, 56 (1991) 471-488.

[4] VISURI, O., et al., Investigation of drag models in CFD modeling and comparison to experiments of liquid–solid fluidized systems, Int. J. Miner. Process 104 (2012) 58-70.

ID#3
URANIUM OXIDE POWDER STUDIES FOR NUCLEAR FUEL FABRICATION

A-C. ROBISSON, N. BLANC, X. ILTIS, C. DUGUAY, C. ABLITZER
CEA, DES.IRESNE.DEC,
Cadarache, France

Abstract

This paper provides a comprehensive overview of R&D studies carried out at the French Alternative Energies and Atomic Energy Commission (CEA) to understand the processes involved in nuclear fuel fabrication. Through coupled experimentation and simulation studies, the ongoing work aims at the complete simulation of all operations within a fabrication process with the objective of predicting the behaviour of powders and the characteristics of the final pellets. In this context, the CEA has developed a multiscale experimental programme focused on the flowability of UO_2 powders, involving data acquisition from the microscopic scale to the macroscopic one. Thus, starting from parameters characterizing a powder particle, laws are sought to predict behaviour at the powder bed scale or the scale of a process operation. To achieve this, new methods are under development, particularly concerning the description of the shape and mechanical properties of UO_2 particles, followed by the measurement of powder flow properties. The data are interpreted using two different approaches, one based on the calculation of inter-particle forces, and the other based on statistical correlations. These promising methods are coupled with instrumentation studies of specific process operations to acquire continuous data. The entirety of these experimental data contributes to feeding the simulation studies also conducted at the CEA.

1. INTRODUCTION

Granular materials are used for nuclear fuel fabrication considering the following main operations: mixing, milling, filling, conveying, pressing, and sintering. Most of these steps require that the powders involved exhibit favourable flow characteristics. However, powders are known to behave unpredictably, leading to various problems such as pipe clogging in case of poor flowability, agglomeration or segregation. Up to now, although important variations between nuclear powders have been noticed, the link between the powder characteristics and their flowability remains unclear. This problem, common to many industries using powders, has been the subject of numerous publications in the fields of agri-food and pharmacy. This becomes even more crucial in the nuclear domain, as clogging phenomena in pipes or devices may require human intervention, thereby potentially leading to an increase in the doses received by workers.

In this context, the CEA has built a coupled experiment/simulation research programme in order to predict the behaviour of the granular material UO_2 and/or PuO_2 in a nuclear fuel fabrication process. This approach, initiated several years ago at the CEA, aims at establishing predictive simulations of the whole process. These predictive laws could have several objectives: improving fabrication process performance, establishing laws to scale up granular material behaviour from the laboratory scale to the industrial one, and preventing changes induced by a modification of feeding powders. In practice, in addition to the development of numerical methods for simulating granular media, this coupled approach requires experimental studies to deepen our understanding of rheological properties of UO_2 and PuO_2 powders, to acquire input data to feed and calibrate simulations or to validate simulation results [1].

Although the CEA is committed to a simulation approach of all the steps of nuclear fuel fabrication process (milling, pressing, sintering, etc.), the work presented in this document focuses on powder flowability studies during powder transportation between all these different steps. For this reason, nuclear powder flow properties have been studied at CEA using different methods. These studies are based on experimental works on actinides powder surrogates and on milled or mixed UO_2 and PuO_2 powders [2-4]. More precisely, three R&D axes on UO_2 powders are undergoing in our laboratory:

- Studies on the description of a powder agglomerate: physical variables, particle shape factors, microstructure and mechanical resistance;
- Studies on the development of empirical or semi-empirical models to describe powder flowability from powder characteristics;
- Instrumentation studies: live monitoring of the powder behaviour in a device: a study case of a rotating drum.

2. POWDER DESCRIPTION

2.1 Interparticular forces

For a powder, the flowability corresponds to its capacity to flow easily, in a regular and constant way, and as individual particles [5]. Powders experience different forces: gravitational force and inter-particular cohesive forces. In order to quantify the importance of forces between 2 particles, the individual Bond number has been defined as the ratio between cohesive forces and the weight of the particle [6]. If the value is greater than 1, it means that cohesive forces predominate, if it is less than 1, the particle weight predominates which promotes free flowing. By extension, a population dependent granular Bond number can be defined for the whole size distribution of a powder constituted of an N particle size classes [6].

$$Bo_G = \left(\sum_{k=1}^{N} \sum_{l=1}^{N} \frac{f_s(d_k).f_s(d_l)}{Bo_{g,kl}} \right)^{-1} \quad (1)$$

with $f_s(x_k)$ and $f_s(x_l)$ the surface distribution functions according to the particle sizes $d_k$and d_l respectively (considering the particles as spherical), $Bo_{g,kl}$ the individual Bond number between particles of $k$and l calculated as:

$$Bo_{g,kl} = \frac{F_{kl}}{W_{p,kl}} \quad (2)$$

where F_{kl} the cohesive forces between two spherical particles of size $d_k$and d_l, $W_{p,kl} = \frac{\pi}{6} g \rho_s \sqrt{{d_k}^3 {d_l}^3}$ the geometric mean of the weight of particles k and l; ρ_s the density of the particles.

For dry powder, assuming that the electrostatic and capillary forces can be neglected, it is generally considered that inter-particular forces are equal to the van der Waals forces. The modified Rumpf equation is used to calculate the van der Waals forces [7]:

$$F_{kl} = \frac{H}{12{z_0}^2} \left[\frac{\hat{d}_{kl}}{2\left(1 + \frac{d_{asp}}{2z_0}\right)^2} + \frac{3\hat{d}_{asp}\hat{d}_{kl}}{\hat{d}_{asp} + \hat{d}_{kl}} \right] \quad (3)$$

with $H = 3.4 \times 10^{-1}$ J the Hamaker constant of UO_2, $z_0 = 0.4$ nm [8] the distance between two spherical particles in close contact, $\hat{d}_{kl}$ the harmonic mean diameter for the particles k and l, $\hat{d}_{asp}$ the harmonic mean diameter for the asperities at the particle surface, often approximated to 200 nm [9].

2.2 Studied powders

Equations for modelling inter-particular forces put forward the importance of the particle physical characteristics in light of the powder flow properties. In practice, the quantities related to particle density, size, and surface roughness can govern the flow behaviour of powders.

Thus, in the initial stage, the experimental approach adopted is based mainly on the physical characterization of different uranium oxide powder batches. Indeed, whatever the nuclear fuel is considered (MOX or UO_X), the UO_2 powder corresponds to the major compound used in the fabrication process. In this context, the studied powders correspond to UO_2 powders obtained by different manufacturing steps (milling, sieving, stirring, lubrication, etc.) but also to UO_2 model powders whose characteristics have been simplified in relation to the first powders. It is to be noted that all these powders were prepared for R&D purposes, so the aforementioned process operations were carried out at the laboratory scale. Some studies are also performed with inactive surrogate powders such as alumina and glass powders in order to have a wider range of characteristics and to work on model particles.

2.2.1 *General powder description*

Studies performed at the CEA to describe UO_2 powder agglomerates (also named particles in this document) concern:

- Their microstructure made up of distinct granular entities: crystallites, aggregates, and agglomerates. This first level of qualitative analysis is carried out by SEM: an example is given in Figure 1;
- Their particle size distribution, measured by laser granulometry after liquid or dry dispersion, with parameters of interest such as characteristic diameters and distribution width (Figure 2);
- The agglomerate density determined by mercury porosimetry;
- Their morphological distribution measured by image analysis taking into account morphological descriptors such as elongation, circularity, and solidity;
- Their mechanical properties determined by micro-compression tests;
- Their surface rugosity whose measurement and definition of the relevant quantities are still under investigation by image analysis or atomic force microscopy.

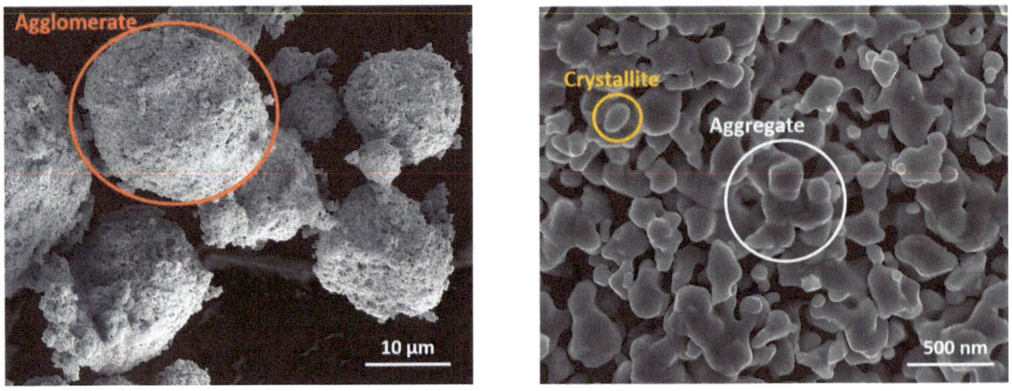

FIG. 1. Example of a microstructure of an agglomerate of UO_2 powder, observed by SEM at two different magnifications.

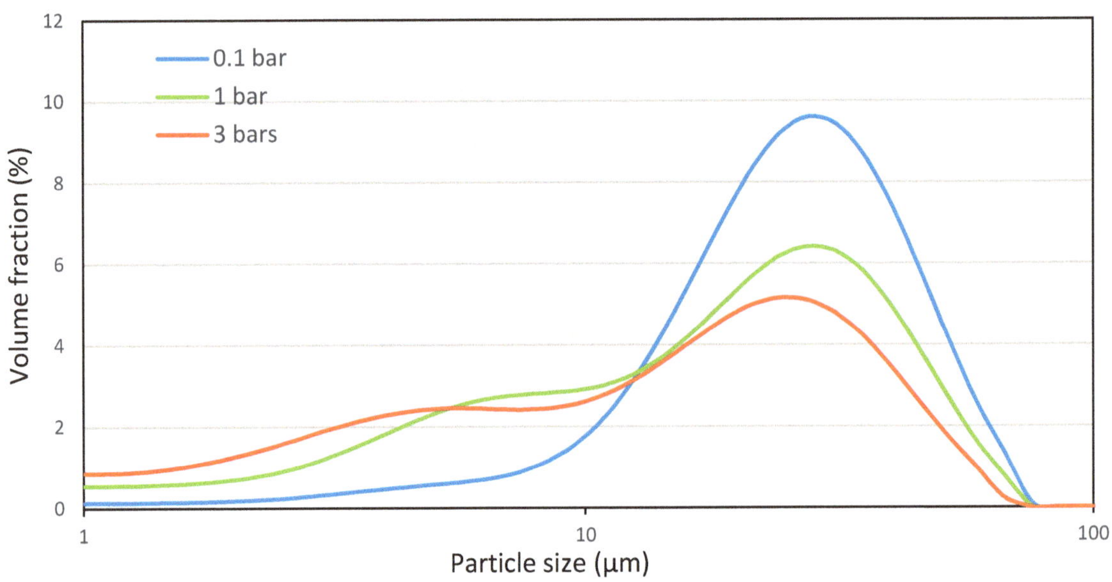

FIG. 2. Example of a particle size distribution for a UO_2 powder after dry dispersion (measurements performed for 3 dispersion pressures: 0.1, 1, and 3 bars).

While these physical characteristics depend on the initial synthesis process of the powder, the different following steps of the fabrication process also influence them. In this publication, particular attention will be paid to methods that are absent or sparsely represented in the existing literature for UO_2 agglomerate description i.e. the measurement of the morphological distribution and the mechanical properties.

2.2.2 Agglomerate shape description

Despite the preponderant influence of the particle size on the powder flowability, particle morphology plays an important role in the case of powders with a similar size distribution. The morphological distribution of UO_2 particles is characterized by 3 shape parameters of interest: elongation *e*, circularity *c*, and solidity *s* (Table 1).

TABLE 1. SHAPE PARAMETERS USED FOR UO_2 PARTICLES CHARACTERIZATION

Shape parameter	Definition	Graphic representation	Remarks
Elongation	$e = 1 - \frac{d_{Feret,min}}{d_{Feret,max}}$	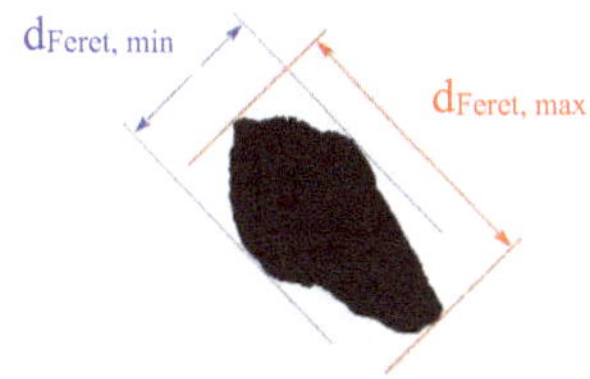	Characterizes elongation and the overall shape of the particle
Circularity	$c = \sqrt{4\pi A/P^2}$		Characterizes the overall shape of the particle and the angularity of its contours.
Solidity	$s = \frac{A}{A_{Env.Conv.}}$	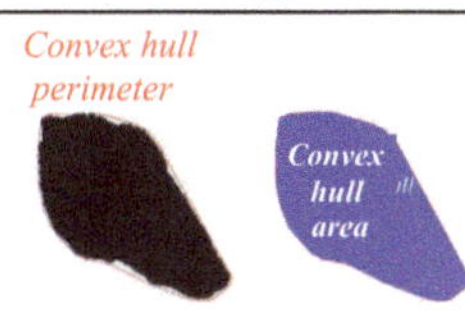	Characterizes the angularity and irregularity of the contours of the particle.

Two types of techniques were used for acquiring these distributive quantities:

- The image analysis of the powder coated with resin. After hardening, the sample is polished to observe sections of particles. This method allows to easily individualize particles and perform a suitable image treatment on microscopy micrographs (with the ImageJ software) [10]. It is well adapted to particles whose global shape is not modified by a random section. Typically, a particle with a small solidity value could be considered as several separate particles in function of the position and the orientation of the section;
- A determination with a morpho-granulometer from Occhio Instruments.

The characterization of powders obtained by different fabrication steps shows the impact of these operations on the morphological quantities. More precisely, brewing improves the particle circularity (Figure 3) and lubrication with 0.3% mass of zinc stearate leads to a significant decrease in their elongation (Figure 4).

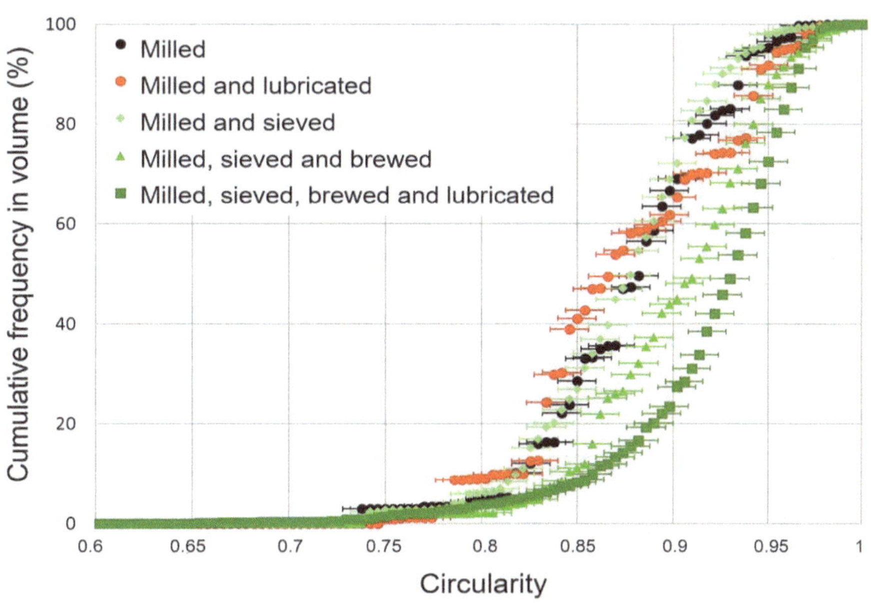

FIG. 3. Impact of the operation process on UO_2 particles circularity (obtained from image analysis).

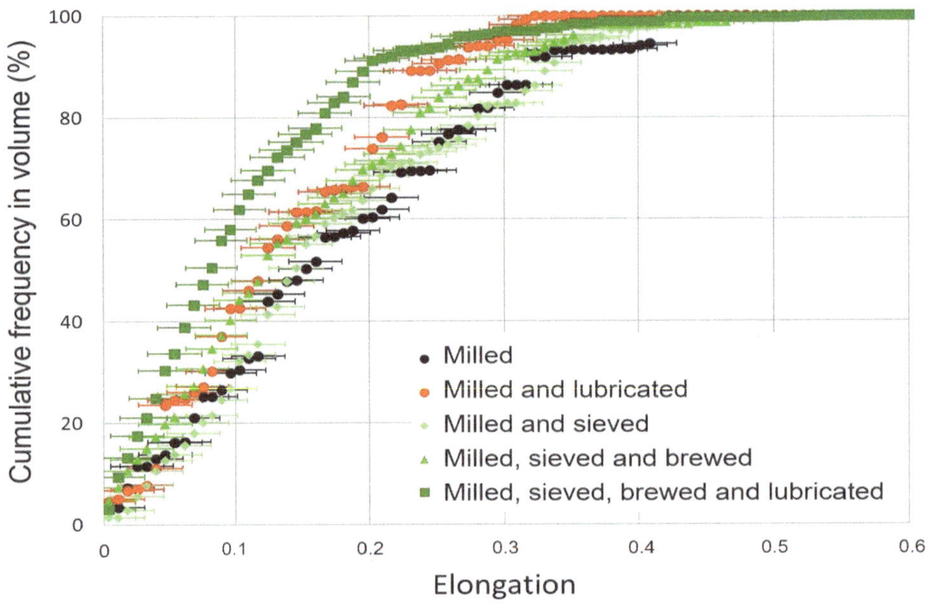

FIG. 4. Impact of the operation process on UO_2 particles elongation (obtained from image analysis).

2.2.3. *Agglomerate mechanical properties*

To understand better the powder evolution during a process, the mechanical resistance of an agglomerate needs also to be known. In fact, depending on the stress state of the powder during the different steps (milling, mixing, sieving, conveying), agglomerates can be more or less consolidated. This consolidation state can have an impact on the particle fragmentation resistance. Thus, in function of the particle consolidation state due to the process, particle size distribution can evolve during the process. These changes can have a major impact on the powder flowability.

In this context, experiments are performed to quantify the breaking strength of UO_2 agglomerates by compression tests in an SEM (Figure 5). For this study, we have selected particles with a size in the range of 200–315 µm obtained after 3 different process steps (Table 2). This range corresponds to the major granulometric class of the studied samples. For one of the samples, different particle sizes were analysed. An example of compression is given in Figure 5 and evidences four steps:

- A: Beginning of the compression;
- B: First break;
- C: Secondary breaks;
- D: Breaks propagation.

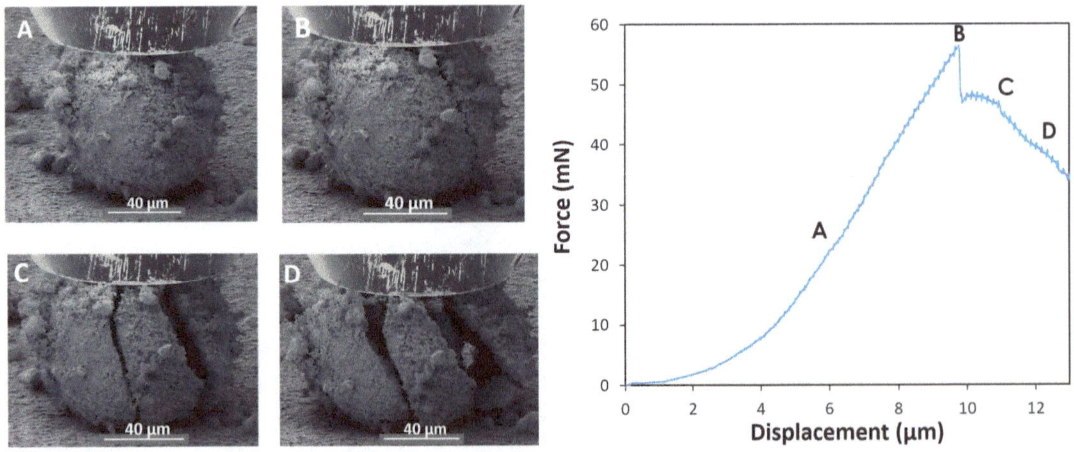

FIG. 5. Compression of an UO_2 agglomerate with a micro-indenter in a SEM.

TABLE 2. SAMPLES ANALYZED BY COMPRESSION TESTS

UO_2 powder	Sieved	Sieved and brewed	Sieved, brewed, and lubricated			
Size range of the analysed particles (µm)	200-315	200-315	0-45	45-100	100-200	200-315
Number of analysed particles	28	27	27	29	32	19

The results of the compression tests allowed the determination of the breaking stress (B point) for each particle analysed. These data are processed using Weibull statistics for the determination of the Weibull constant and modulus (Figure 6). The results show a lower mechanical strength of the only sieved powder compared to the other ones. It seems that the brewing and lubrication operations allow UO_2 agglomerates to consolidate. However, these values tend to indicate that lubrication has no impact on this mechanical resistance. In addition, in the three tested cases, the Weibull moduli are in good agreement with the breaking of fragile materials.

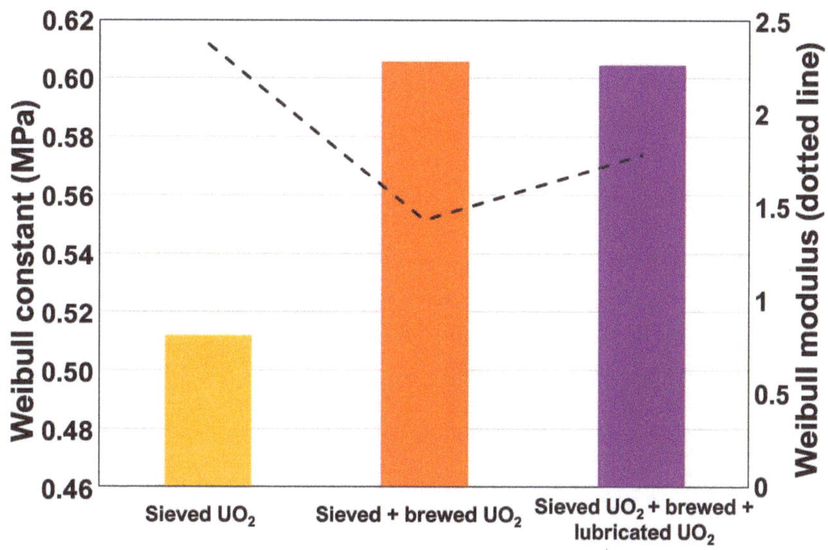

FIG. 6. Results of compression tests on 3 types of UO_2 powder.

Moreover, results obtained on sieved, brewed, and lubricated UO_2 powder showed that tensile strength is linked to the particle size, probably owing to different particle microstructures that can exist between fine and coarse particles. More precisely, for a milled powder, agglomerates under 45 µm present a better mechanical resistance than agglomerates over 45 µm (Figure 7).

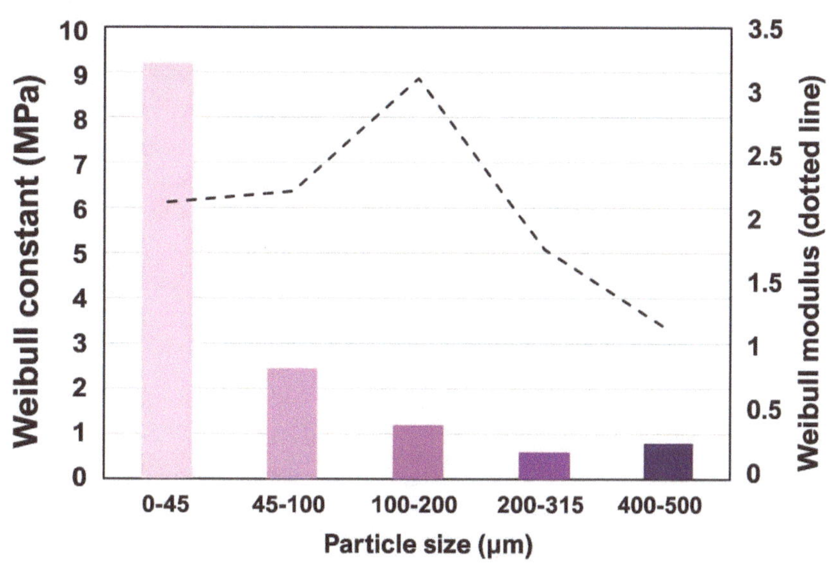

FIG. 7. Results of compression tests as a function of the particle size (sieved + brewed + lubricated powder).

3. POWDER RHEOLOGICAL BEHAVIOUR

3.1 Powder flow properties

There are several flow properties to describe the behaviour of a powder in different situations: static-dynamic transition (flow initiation), dynamic flowing, and powder bed rearrangement. Two different approaches were used for measuring these properties: first level analyses via the determination of apparent and tapped densities and characterization using a powder rheometer through shearing, compressibility, and permeability tests.

Furthermore, three different situations of flowing have been considered in this work:

- Flow initiation:

Shear tests with a rheometer allow us to quantify the powder transition between a static and a dynamic state. This property is characterized by several quantities such as the cohesion stress, the flow index, and the effective internal friction angle. In this study, shear tests are performed with an FT4 rheometer.

- Dynamic flow:

This kind of flow corresponds to an established flow rate. It can be characterized through a test named variable flow rate, performed with an FT4 rheometer, including the determination of the Flow Rate Index. A rotating drum can also be used for measuring this property, allowing the determination of the angle of repose or the avalanche energy.

- Powder bed properties:

There is a wide range of powder bed properties. The Carr index I_c and the Hausner ratio R_H lead to a first level description of the powder bed ability to rearrange itself. These indexes are defined from apparent and tapped densities [11]. Similarly, the compressibility test measures the apparent volume variation of a powder bed under a given applied stress. Furthermore, the permeability can be determined to evaluate the capacity of a powder bed to allow the passage of a fluid through its pores or structure.

3.2 Prediction of the powder rheological behaviour

In addition to phenomenological considerations on the interpretation of UO_2 powder flow properties, two different approaches are being pursued to better predict the powder behaviour from its physical characteristics: a semi-empirical approach based on the search for correlations between the rheological properties and the Bond number and a statistical approach.

3.2.1 Semi-empirical model

In the literature, several authors have demonstrated the existence of a power law to link the flow index ff_c with the population dependent granular Bond number Bo_G. Their studies were performed on organic powders (pharmaceutic powders) or inorganic powders [6, 12, 13, 14]:

$$ff_c = \alpha {Bo_G}^{-\beta} \tag{4}$$

In our work, several powder batches were analysed in order to verify whether this power law also applies to actinide oxide powders. For that purpose, model powders allow to work on samples with a narrow particle size distribution compared to industrial powders and with similar morphologic distributions. Our results show that a power law can indeed be deduced for cohesive powders with a Bond number greater than 1. Moreover, the obtained β coefficient is in very good agreement with those mentioned in the literature for inactive powders (Figure 8). Concerning non-cohesive powders, typically with a size greater than 1000 µm, van der Waals forces are no longer predominant, and the flow is governed by particle weight, friction forces, or electrostatic forces.

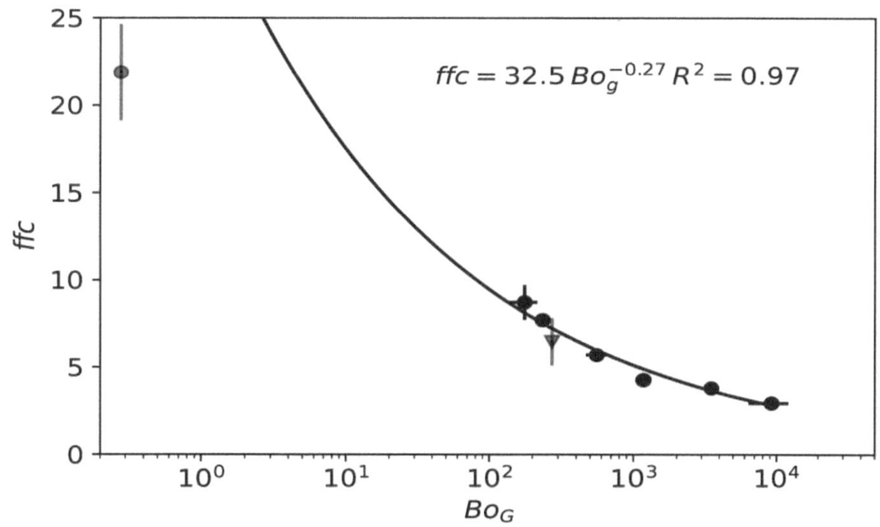

FIG. 8. Flow index ff_c as a function of the granular Bond number for different UO_2 samples.

Some works in the literature have tried to interpret the β coefficient from the theoretical Rumpf's equation, employed to describe the relationship between the tensile strength of a powder bed and the cohesive interparticular forces [15]. However, this interpretation could not be validated in the case of this study. Indeed, assuming the tensile strength of the powder bed is approximated to the cohesion, a coefficient equal to -0.59 is obtained, this value being greater than that adjusted to the experimental data (-0.27).

Ongoing studies aim at better understanding this power law including the study of the impact of the particle morphology and rugosity.

3.2.2 Statistical approach

A statistical approach has begun with the goal of putting forward the potential links between some powder characteristics and some flow properties. To do so, two methodologies using linear relationships were tested:

- A multiple regression linear;
- A correlation matrix using the Pearson correlation coefficients.

The first method aims at establishing a mathematical correlation between independent variables and a dependent variable with the objective of predicting the dependent variable based on the independent ones [16]. In the present study, the physical attributes of the UO_2 powder are treated as independent variables within the framework of multiple linear regression. Specifically, the independent variables taken into account are the characteristic diameters of the particles, the width of the particle size range, and morphologic descriptors [4]. Concurrently, the rheological properties are designated as the dependent variables.

Thus, linear regression models were determined for the flow index, the internal friction angle, the flow rate index, the permeability, the Carr index, and the compressibility. Some relationships are given below as an illustration:

$$ff_c = 3.80 + 0.41\, d_{10} - 1.09\, Span - 0.74\, SF_{50,\,elongation} \quad (5a)$$

$$Flow\,Rate\,Index\,RI = 1.20 - 0.09\, d_{10} + 0.09\, Span + 0.11\, SF_{50,\,elongation} \quad (5b)$$

$$Carr\,Index = 18.07 - 1.84\, d10 + 1.89\, Span + 1.56\, SF_{50,\,elongation} \quad (5c)$$

where d_{10} represents the fraction of fine particles, Span corresponds to the width of the particle size range and $SF_{50,\,elongation}$ and $SF_{50,\,circularity}$ represent median values of particle elongation and circularity. Figure 9 shows the distribution of values predicted by these models as a function of experimental values (error bars represent the 95% confidence interval).

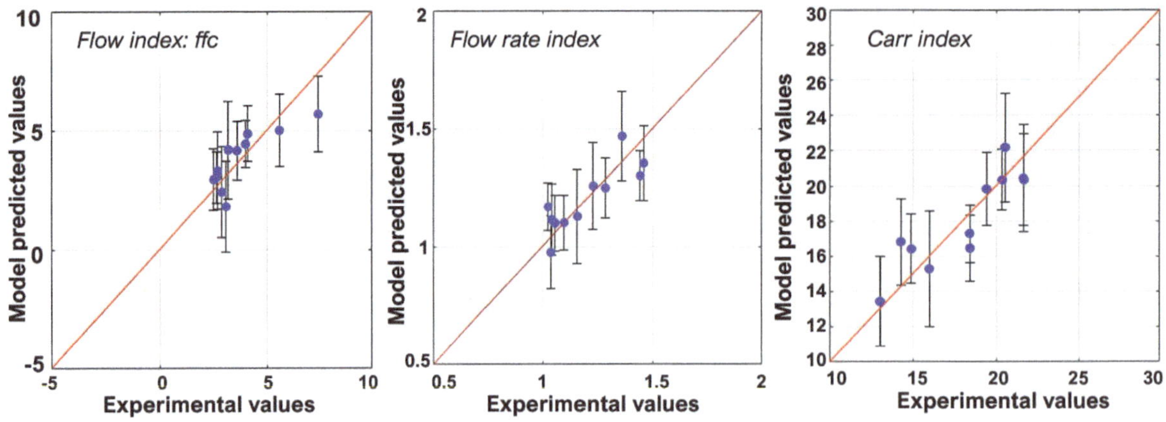

FIG. 9. Model-predicted values as a function of experimental values for the flow index, the flow rate index and the Carr index.

All these results highlight the major influence on flow properties of three characteristics: the fraction of fines, the width of the particle size range, and the particle elongation. Particle morphology, and in particular elongation, appears to be one of the most important parameters influencing powder flowability.

The second method usually requires a large number of samples. This methodology for studying rheological properties has not yet been widely reported in the literature [17-18]. Studies are currently underway to obtain a large number of characteristics and properties using inactive powders. Only a few preliminary results on UO_2 powders are presented in this document. For this statistical approach, FT4 powder rheometer measurements are correlated to granulometric data for 8 UO_2 powders with similar morphologic characteristics.

As shown in Figure 10, using the logarithms of these data, possible correlations in the form of power-laws are obtained. The absolute values of the Pearson correlation matrix thus enable observation that the flow index presents a high-power law correlation with the fine fraction and the span (Figure 11). Such is also evident in the case of compressibility; however, the correlation between permeability and granulometric quantities is less pronounced.

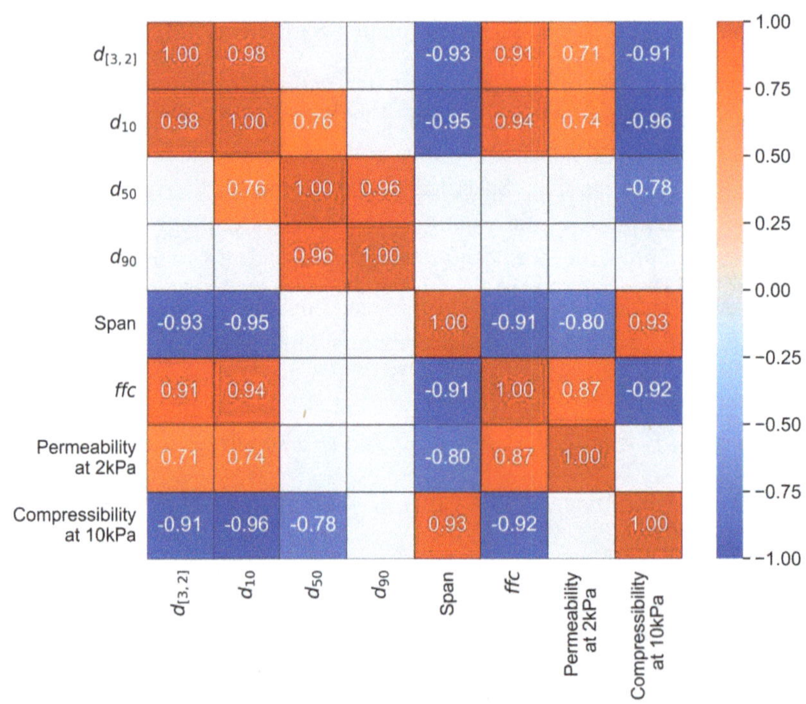

FIG. 10. Absolute values of the Pearson correlation matrix of logarithms for granulometric quantities ($d_{[3,2]}$, d_{10}, d_{50}, d_{90}, Span) and flow properties (ffc, permeability, compressibility).

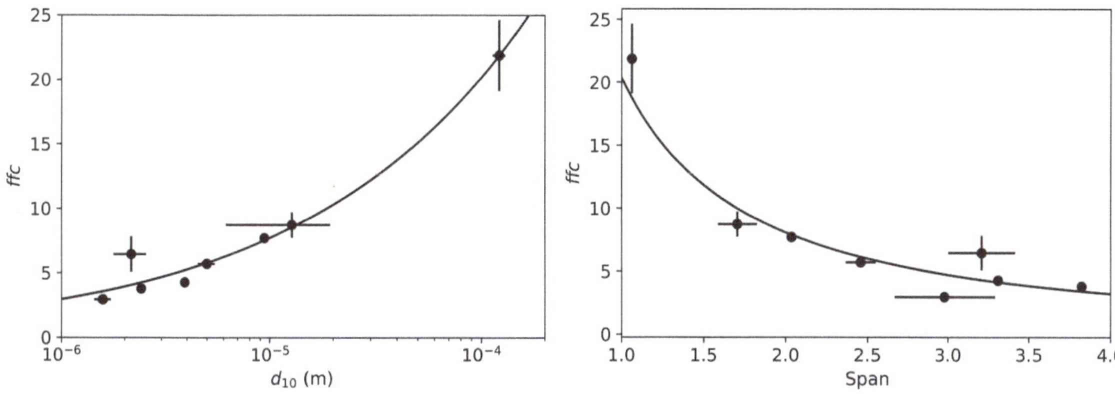

FIG. 11. Correlations concerning the flow index, data interpolated using a power-law function.

4. INSTRUMENTATION APPROACH: THE BALL MILLING EXAMPLE

Work on fuel fabrication at CEA also includes the instrumentation of some equipment used in the processes. The objectives are the following:

- Closely monitoring a procedural operation to deduce the continuous evolution of the powder (or pellets) characteristics;
- Continuously adjusting the implementation conditions of a process fabrication in order to improve it but also to early detect unexpected changes for better prevention of potential issues;
- Acquiring a large amount of data to feed the simulation. In particular, data assimilation methods need a large number of values derived from experimentation to be relevant.

In this context, a comprehensive study has been initiated involving experimentation on UO_2, instrumentation of a device, and simulation using the discrete element method and data assimilation methods. The selected operation for this work is the ball mill, which is traditionally used for MOX fuel production, to micronize and homogenize the distribution of uranium and plutonium.

The ball milling involves a rotating drum containing both the powder and the balls. The balls correspond to metallic orthocylinders (U-Ti alloy in the case of the MOX). The mechanisms involved in the drum are particle fragmentation, attrition, agglomeration, or even compaction. There are different regimes of the movement inside the drum in the function of the charge (powder+balls), the drum size, and the rotation speed: sliding, rolling, cataracting, cascading, and centrifugation (Figure 12).

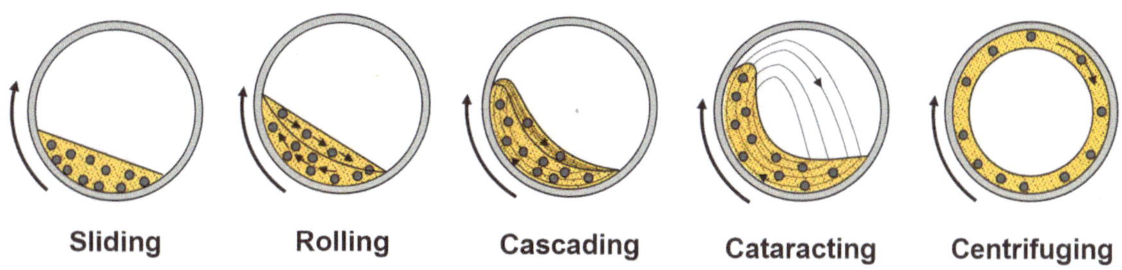

FIG. 12. Different ball milling regimes.

The instrumentation of a ball mill corresponds to online data acquisition from several sensor types:

- Acoustic and vibrational signals (microphones, accelerometer, laser vibrometry, etc.);
- Image analysis (speed fields, free-surface profile, angles, etc.);
- Device data (rotation speed, power torque, etc.).

Initially, tests are conducted on inactive simulant powders to establish acquisition methods, which are subsequently applied in a second phase to a ball milling in a glove box on UO_2 powder. An example is given here with the use of a rapid camera to acquire images on a transparent side wall of the drum.

Thus, by employing raw images capturing the motion of powder and grinding bodies within the mill, image processing using particle image velocimetry facilitates the analysis of the charge profile and barycentre (Figure 13). Subsequently, the profile enables the examination of the milling regime, which may vary depending on the progress of the milling (linked to the powder granulometry). These images can also be used to investigate the homogenization of two differently coloured components, determining a mixing homogeneity degree through specific methods. The experimental results are then compared to simulation data.

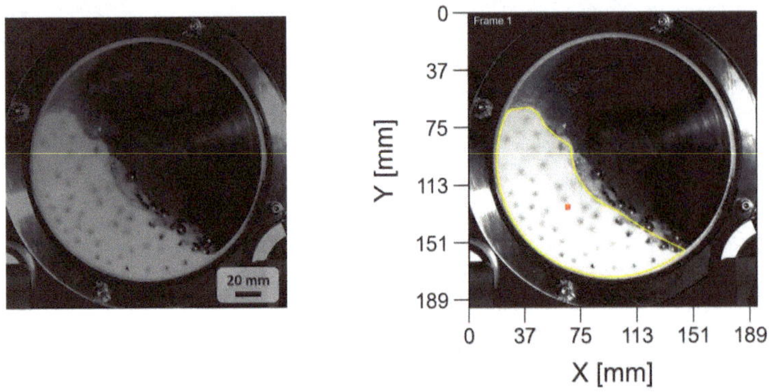

FIG. 13. Raw data (left), flow profile (in yellow) and barycentre (in red) determined by image analysis (right).

These acquisitions also make it possible to determine velocity fields (Figure 14). These data will soon be available for comparison with the values obtained by the discrete element method.

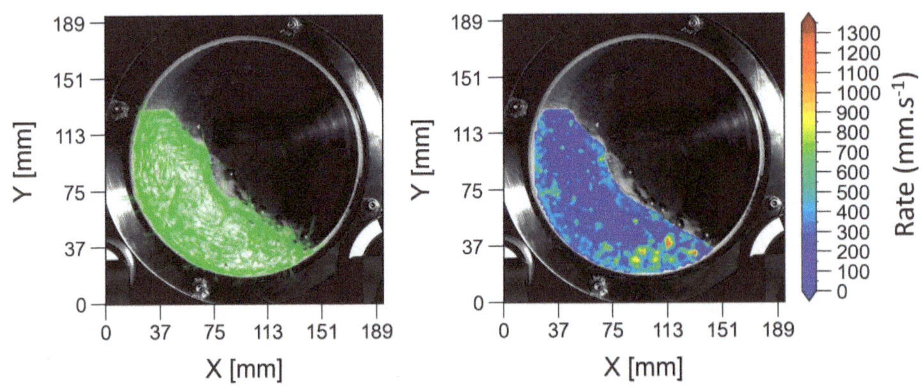

FIG. 14. Velocity vector fields (left) and velocity mapping (right) determined from experiments.

5. CONCLUSION

The work presented in this publication provides an overview of the approach adopted in recent years at the CEA regarding studies on fuel manufacturing. The objective of understanding the physical phenomena within a manufacturing process is inherent to the initial goal of predicting the behaviour of the powder in each stage with a view to optimization. Beyond the stages of transport and milling, the overall approach of coupling experimentation/simulation is also adopted for the pressing and sintering stages. The ultimate goal is to build a multi-D, multiphysics, and multiscale simulation tool for the fabrication process using forecasting to build a multi-D, multiphysics, and multiscale simulation tool for the fabrication process using forecasting models for all the stages.

REFERENCES

[1] VU., D-C., Quasi-static and dynamic granular flows : scaling behavior, microstructure, and particle shape effects, PhD of l'Université de Montpellier (2023).

[2] MADIAN, A., Étude de l'influence des caractéristiques physiques des poudres UO2 sur leurs propriétés rhéologiques, PhD Université de Technologie de Compiègne (2019).

[3] GIRAUD., M., Analyse du comportement rhéologique des poudres à partir des propriétés des grains, application à l'étude d'un procédé de broyage/mélange pour la préparation du combustible nucléaire MOX, PhD of Toulouse University (2020).

[4] BEAUNAC., E., Étude du lien entre les caractéristiques physico-chimiques des poudres et leur comportement lors des phases de transfert, PhD Université de Technologie de Compiègne (2021).

[5] SALEH., K., GUIGON., P., Mise en œuvre des poudres - Granulation humide : bases et théorie" Techniques de l'Ingénieur, Référence J2253, (2009).

[6] CAPECE., et al., Prediction of powder flow performance using a multi-component granular Bond number, Powder Technol. **286** (2015) 561-571.

[7] CHEN., Y., et al., Fluidization of coated group C powders, AIChE J. **54** (2008) 104-121.

[8] ISRAELACHVILI., J. N., Intermolecular and Surface Forces, Elsevier, Burlington, Massachusetts, 2011.

[9] L. MASSIMILLA., L., DONSI., G., Cohesive Forces Between Particles of Fluid-Bed Catalysts, Powder Technol. **15** (1976) 253.

[10] D'ANGELO., C., et al., Quantification of the morphology and roughness of oxide powder particles in relation to their manufacturing history and flow properties, PARTEC, Nüremberg, September 2023.

[11] LETURIA., M., et al., Characterization of flow properties of cohesive powders: A comparative study of traditional and new testing methods, Powder Technol. **253** (2014) 406-423.

[12] CAPECE., M., et al., On the relationship of inter-particle cohesiveness and bulk powder behavior: Flowability of pharmaceutical powders, International J. Pharm. **511** (2016) 178-189.

[13] BERNARD-GRANGER., G., et al., Rheological properties of alumina powder mixtures investigated using shear tests, Powder Technol. **345** (2019) 300-310.

[14] KUNNATH., et al., Selection of Silica Type and Amount for Flowability Enhancements via Dry Coating: Contact Mechanics Based Predictive Approach, Pharm. Res. juill. (2023).

[15] GIRAUD M., et al., Investigation of a granular Bond number based rheological model for polydispersed particulate systems, Chem. Eng. Sci. **228** (2020) 115971.

[16] FROST., J., Regression Analysis: An Intuitive Guide for Using and Interpreting Linear Models, Ebook statisticsbyjim.com (2019).

[17] HEJDUK., A., et al., Impact of co-processed excipient particles solidity and circularity on critical quality attributes of orodispersible minitablets, Powder Technol. **387** (2021) 494.

[18] SHIER., A. P., et al., Development of a predictive model for gravimetric powder feeding from an API-rich materials properties library, Int. J. Pharm. **625** (2022) 122071.

ID#4

MOX FUEL PROCESS: POWDERS AND PELLETIZING FABRICATION EXPERIENCE

G. KERBOUL, J-M. MARIN
Orano,
France

Abstract

The paper provides basic descriptions of the different steps of manufacturing MOX pellets at ORANO Melox Facility in France. It gives an overview of the so-called MIMAS process used at Melox and presents the main equipment used in the preparation of powders until pressing. The main feedback from this equipment since the start of Melox is also discussed.

1. INTRODUCTION

The MELOX plant is, to date, the only industrial manufacturing plant of LWR MOX operating worldwide. With nearly 3000 tHM of MOX fuel produced since its commissioning in 1995, it has accumulated a unique track record and produced LWR MOX for France and various foreign customers.

Melox manufacturing process is based on the MIMAS process - MIcronisation MASterblend, which is detailed below.

2. MELOX MOX FUEL POWDER PROCESS

2.1 Overview of the MIMAS process

When we talk about MOX fuel powder process at Melox, we are dealing with the so-called MIMAS process, meaning MIcronization of a MASter blend. This is a two-step process to produce a high plutonium content master blend that will be diluted with a uranium dioxide powder in order to manufacture (U, Pu)O_2 fuel pellets, in the range of 2 wt% to 12 wt% of Plutonium content. UO_2 powders from a wet root process or a dry root process can be used in the MIMAS process. The MIMAS process also allows high recycling ratios. Indeed, recycling scraps produced during the fabrication process is a necessary step and the MIMAS process is capable of recycling up to 15 wt% of recycled materials.

The different steps of the MOX fuel manufacturing at Melox are shown in Figure 1.

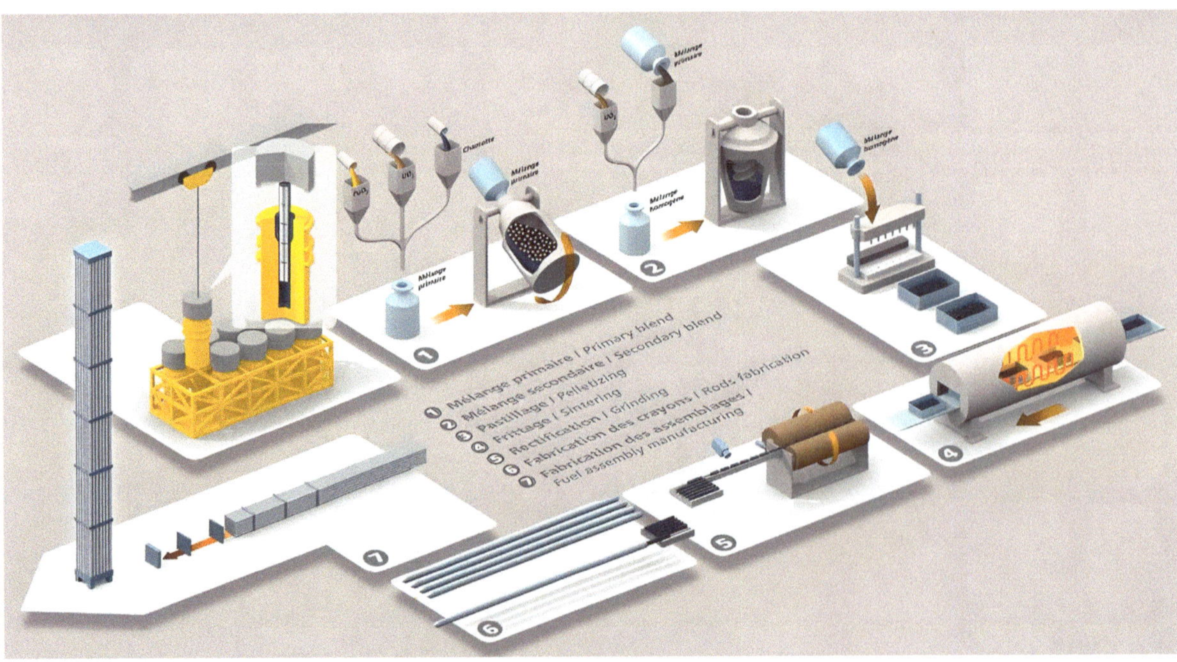

FIG. 1. The MOX MIMAS process at Melox.

The plutonium is shipped from the reprocessing facilities at La Hague to the MOX fuel fabrication plant, unloaded, and stored. The plutonium oxide is dosed and mixed with uranium oxide and recycled scrap powders to form the primary blend (see process step 1, Figure 1). This primary blend is milled, sieved, and diluted with uranium oxide powder (see process step 2, Figure 1). During this secondary blend, the plutonium content is adjusted according to the specifications of each customer. The MOX powder is then mixed with additives and pelletized with hydraulic presses (see process step 3, Figure 1). The MOX green pellets are sintered at around 1700°C (see process step 4, Figure 1), to achieve the specified density and physical and chemical characteristics.

The sintering operation is followed by the grinding operation (see step 5, Figure 1). For safety reasons, no water is allowed in the gloveboxes. So, MOX pellets are "dry" ground, using a centreless grinding machine. This operation leads to produce cylindrical pellets with the specified diameter.

From this stage, the process is very similar to that of UO_2 fuel fabrication plants, i.e. introduction of the column of pellets into the fuel rods (see process step 6, Figure 1); pressurisation of fuel rods using helium; welding of the end plug; QC of the rods and MOX fuel rod assemblies (see process step 7, Figure 1).

2.2 Powder Process Equipment feedback

Plutonium oxide powder, uranium oxide powder, and recycled scrap powder are dosed and mixed to form the master blend of the MIMAS process. The plutonium content is at this stage around 20 wt% minimum. This primary blend is milled, using a very energetic miller, in order to have a perfect micro-homogeneity between the powders. The powder is then forced-sieved with a ~250 µm mesh sieve to control the particle size distribution.

Then the sieved primary blend is dosed with UO_2 powder to achieve the specified plutonium content, less than 12 wt%. Several of these secondary blends are homogenized together. A pore former agent is added during the homogenisation step to allow better control of the sintered density. To facilitate the pressing operation, a lubricant agent is mixed with the homogenized secondary blend.

In the paragraphs below, the main feedback from some equipment used is given.

2.2.1 Primary blend blender (Figure 2)

The primary blend homogenisation equipment is an Archimedes orbital screw blender which can operate at temperatures between 80 to 200°C. Its volume is approximately 60 litres. Electromagnetic hammers are placed on the blender wall to prevent powder retention.

The powder is introduced in the blender using jars rising to the top of the blender, turned upside down so the powder flows by gravity into the blender. The powder is evacuated by opening a spherical valve at the bottom of the blender connected to a vibrating conveyor, itself connected to a jar.

The blender is surrounded by radiation shields to protect workers from radiation doses and a cooling system to cool the equipment.

Cameras and a window at the top of the blender allow surveillance of the operation inside the blender.

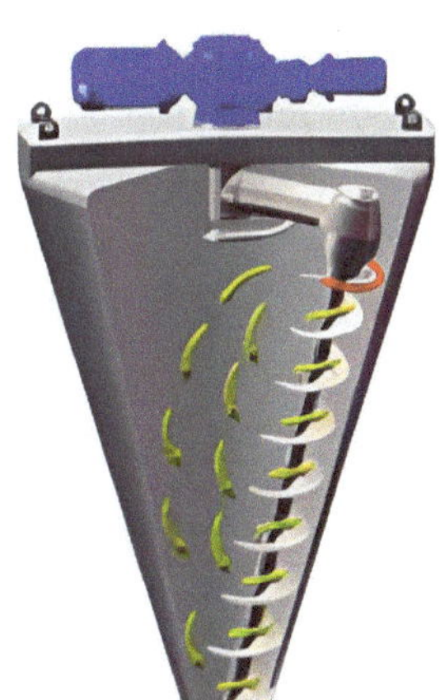

FIG. 2. Primary blend blender.

This equipment has been used for 20 years and no special maintenance events or breakdowns occur. Some equipment surrounding the blender, such as the jar supply device, and the vibrating conveyor, needs to be regularly maintained or changed.

It is to be noted that this equipment has not been used for 6 years. Melox demonstrated that this homogenisation step was no longer necessary due to the efficiency of the following process steps.

2.2.2 Primary blend Ball Mill (Figure 3)

At this process step, four types of different powders are milled together: PuO_2, UO_2, recycled scrap powder, and a lubricant agent. The recycled scrap powder comes from rejected pellets that have been previously milled. The lubricant agent helps the milling operation and avoids an excess of powder agglomeration.

The ball mill is capable of rotating around 2 axes at a speed of several dozens of rotations per minute. The volume of the ball mill inner chamber is approximately 60 litres. The mill is fed using a jar connected to the miller. The ball mill rotates vertically allowing the powder to fall into the inner chamber. The equipment then rotates to a position close to horizontal and is then put into rotation for a certain number of cycles. The milling body consists of uranium metal balls which rotate and fall onto the powder. This type of milling is quite energetic and efficient to obtain good micro-homogeneity between the different powders. The total weight of uranium balls inside the mill chamber is approximately 350 kg. At the end of the milling time, the mill goes back into a vertical position allowing the powder to fall by gravity into the connected jar. Irradiation shields are placed around the mill and the ventilation system limits the temperature increase during the milling operation.

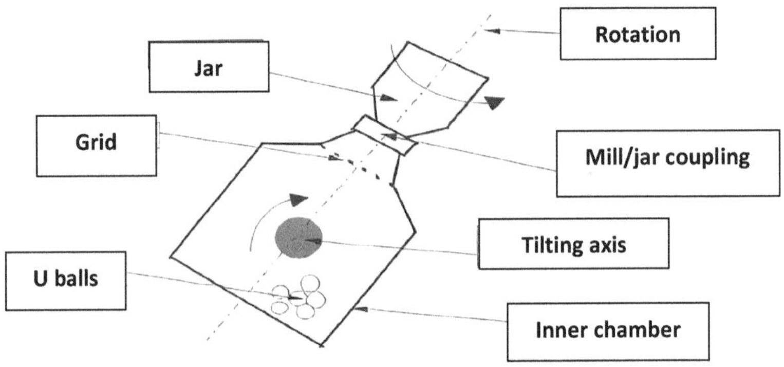

FIG. 3. Primary blend ball mill.

No specific event has been identified for the mill itself. Due to their wear during milling operations, the uranium balls load is adjusted periodically, and after a few years, the full load needs to be renewed.

Maintenance of external equipment is carried out on the weighing devices.

2.2.3 Primary blend forced siever (Figure 4)

To reduce and control the particle size of the primary blend, the primary blend powder is forced sieved.

The equipment used is a hammer mill. Its rotation speed is very high, up to 3000 rotations per minute. The powder which passes through it continuously is projected onto a metallic grid with a mesh size of 200 micrometres.

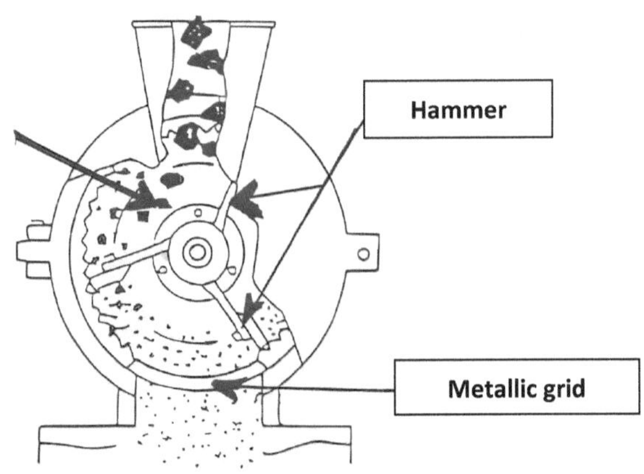

FIG. 4. Hammer mill.

The major manufacturing feedback for 27 years is about the important grid wear. They need to be changed periodically and preventively before any damage occurs. Damage to the grid results in the potential rejection of the powder for quality reasons.

Vibrating conveyors and weighing devices are maintained regularly or changed.

2.2.4 Secondary blend homogeniser

At this step, the secondary blend jars are prepared by adding the fresh uranium powder to the primary blend. The jars are poured into the homogeniser and mixed. A pore former agent can also be added to control the final sintered density, mainly due to the high sintering capability of the uranium powder.

The blender is an Archimedes orbital screw blender with an internal screw, which can operate at temperatures between 80 to 200°C. Its volume is approximatively 320 litres and electromagnetic hammers are placed on the blender wall to prevent the powder retention. The powder is evacuated by opening a spherical valve at the bottom of the blender connected to a vibrating conveyor, itself connected to a jar. Radiation shields are placed around the blender and forced ventilation allows cooling of the system. Cameras and a window at the top of the blender allow surveillance of the operation inside the blender.

In 27 years of operation, the main maintenance was the change of the main motor of the blender, after 20 years of operation, meaning after thousands of tonnes of powder lots.

The other maintenance topics are oil changes in the different mechanical gears to ensure their correct functioning; maintenance on vibrating conveyors and weighing devices; and periodical changes of detectors or sensors.

The homogeneity capability of the blender was demonstrated during the UO_2 performance tests at the beginning of the plant startup through a sampling plan experiment. Sampling at the exit of the homogeniser is no more performed but the homogeneity is still checked on pellets.

2.2.5 Secondary blend lubrification device

The lubrification of the homogenized secondary blend is performed with a ploughshare mixer. Its volume is approximatively 98 litres (~100 kg of powder). Electromagnetic hammers are placed on the lower part of the mixer to ensure proper powder evacuation and prevent retention.

The ploughshare mixer is fed with a jar filled with homogenised secondary blends. The lubricant is poured into the mixer through a pipe connected to a vibrating conveyor. The powder is evacuated by opening the doors at the bottom of the mixer. The powder falls into a hopper connected to a vibrating conveyor and an Archimedes screw to feed the pressing station. Video inspection can be organized on the top of the blender to monitor proper operation. Radiation shields surround the mixer, and the ventilation system allows cooling.

In 27 years still the startup of the plant, maintenance consisted of replacing motors and driving belts, replacing the electromagnetic hammers and different devices below the mixer.

2.2.6 Pressing equipment

Pelletizing is performed using alternative hydraulic presses, a simple effect with floating dies (equivalent to a double effect press). A dynamic compensator facilitates the ejection of the pellets without damage. One press cycle produces 14 pellets at a time. The 14 dies and punches are made in Tungsten Carbide. The pressing equipment is capable of 2.5 cycles per minute. Melox plant has 4 pressing stations.

The major maintenance for 27 years is the change of some motors and sensors, as well as the powder feed capacity surround.

Inspection video can occur thanks to cameras on the top and on the feeding part of the presses.

A periodic green pellet sampling is performed to monitor pellet characteristics (visual aspect, length, density). The pellets are pressed up to 60% of the ceramic theoretical density (around 6–6.5 g/cm^3).

The tungsten carbide tools are replaced according to their wear level.

3. CONCLUSIONS

The Melox process, the so-called MIMAS process was described, and the main powder preparation equipment was detailed. Feedback on the equipment reveals no major maintenance or breakdown for more than 25 years, demonstrating a robust and mature Melox process.

TECHNICAL SESSION II: SINTERING, PELLETIZING, ADVANCED FUEL FABRICATION

ID#5
MODELLING OF THE FUEL SINTERING PROCESS

G. RAVEU*, J. BISCHOFF*, K. KANDEEPAN**, A. COLLIN DE L'HORTET**
* Framatome,
Romans-sur-Isère Cedex

** Framatome,
Lyon

France

Abstract

The sintering in the manufacturing plants within Framatome at Romans-sur-Isère (France), Lingen (Germany) or Richland (USA) uses similar walking beam sintering furnaces for large scale production of nuclear fuel pellets. The sintering process of different products in the same equipment or different equipment present among Framatome is complex with sublimation and reduction reactions involved, but also densification and grain growth, making it difficult to know precisely all the reactions taking place in the different areas of the furnace. Framatome has therefore launched a project to model the furnace and the sintering of the pellets. This modelling has several benefits:

- Perform parametric studies without using the production equipment;
- Increase performance by optimizing the process;
- Compare the different equipment available among Framatome plants and use the best developments;
- Create a teaching tool through the theoretical description of the process and its key influencing parameters.

The modelling is a step-by-step approach, adding new elements or functions at each step. First, the geometry of the Romans-sur-Isère sintering furnace was established with the general modelling of the gas flows and temperatures, then adding more detailed pieces such as the sintering boats or the influence of the walking beam movements. The model created is then validated with experimental data measured during the production campaigns. Parametric sensitivity calculations can be performed to evaluate the influence of different parameters on the process such as gas mixture, and temperatures, and thus ultimately optimize the pellet sintering.

1. INTRODUCTION

Framatome, as an international leader in nuclear energy, designs, develops and manufactures FAs for PWRs, BWRs, research reactors and the next generation of nuclear plants including small modular and advanced reactors. The FA manufacturing line calls upon solid teams of experts and specialists in advanced technologies. It consists of three steps:

- The uranium chemistry for conversion of uranium from UF_6 to UO_2;
- Ceramic fabrication: consisting of compacting the powder into the form of pellets and sintering;
- The mechanical process of manufacturing and assembling complex technical components into FAs.

In this paper, the focus is on pellet fabrication [1], and, in particular, on the sintering step. The sintering is a manufacturing process consisting of a thermal treatment of a powder or compacted powder at a temperature below the compound melting temperature [2-3]. The aim is to increase the mechanical strength of the product. A densification of the compacted powder is observed, accompanied by a shrinkage. The microstructure also evolves with grain growth.

The three nuclear fuel manufacturing plants within Framatome in France, Germany and in the United States of America use similar walking beam furnaces for large scale production of fuel pellets. The sintering process involves complex chemical reactions, including sublimation and reduction, as well as densification and grain growth. This makes it challenging to identify precisely all the reactions taking place in the different areas of the furnace. Framatome has therefore launched a project to model the furnace and the sintering of the pellets. This modelling is part of a transverse and global fuel manufacturing R&D project. The objective is to achieve continuous improvement, optimize fabrication processes, enhance performance, and deliver high-quality products to its customers. Modelling equipment and process is a part of Framatome's approach to master all the parameters and all of the phenomena, which impact or occur during the production of the pellets.

The modelling is conducted through a step-by-step approach, in the following order:

- Create the Computer Assisted Design (CAD) structure of the sintering furnace based on 2D drawings;
- Build a geometrical model by defining all the important parts of the furnace, with associated physical characteristics and interfaces;
- Build a thermal-aeraulic model to define the temperature, gas movements and speeds in the different parts of the furnace;
- Validate the model with experimental data;
- Perform sensitivity calculations and adjust the model.

2. FURNACE DESCRIPTION AND GEOMETRICAL MODEL

A schematic view of a sintering furnace is presented in Figure 1. It a simplified representation that can then admit additions and modifications specific to each furnace (for the different plants) so that they can then be compared.

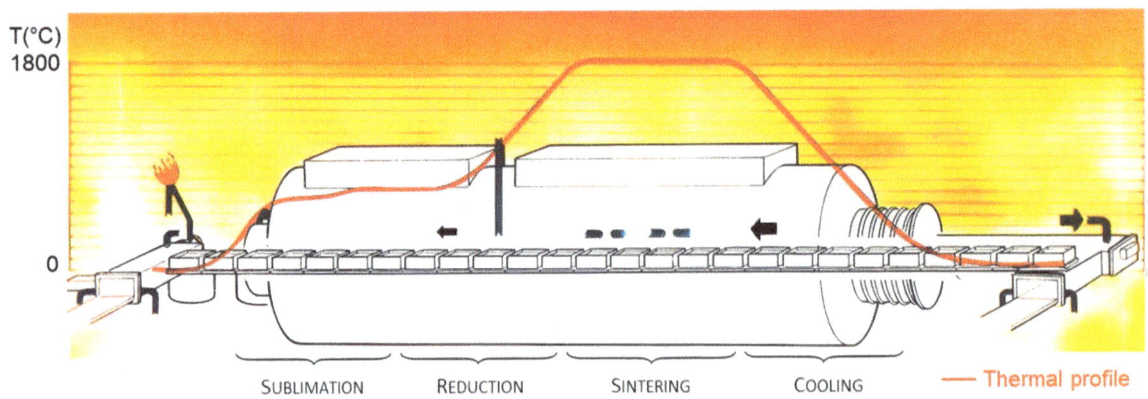

FIG.1. Schematic view of a walking beam furnace.

The sintering furnace is composed of different parts:

- A loading station where the boats full of green pellets enter the furnace through the door;
- An entrance chamber, where the boats are inside the furnace;
- A pre-sintering zone, where the pellets start their thermal treatment beginning with the sublimation of additives and followed by the reduction reaction of U_3O_8 to UO_2;
- A vestibule, a short zone used as a buffer, neither heated nor cooled, located between the pre-sintering and sintering zone;
- A sintering zone, at high temperature up to ~1700–1800°C, where the pellets start their densification and their grain growth;
- A cooling zone with a metallic structure cooled on all sides with water coils, to let boats and pellets cool down before exiting the furnace;
- An unloading station, where the boats come out of the furnace through the door;
- Different layers of refractory bricks all along the inside of the furnace to maintain high temperatures and good insulation;
- An insulating layer and a steel casing envelop all the furnace;
- Chimneys to reject all the excess gases to avoid pressure increase inside the furnace;
- A walking beam mechanism to move boats through the furnace.

The loading and unloading zones of the pellet boats are not considered in this study to focus only on the inside of the furnace where the actual sintering process takes place.

Based on this description and with supporting drawings, the CAD structure of the sintering furnace is built, then the geometrical model of the furnace is created.

3. COMPUTATIONAL FLUID DYNAMICS MODEL

From the geometry presented, a Computational Fluid Dynamics model on STAR-CCM+ v2021.3 is developed to simulate the walking-beam sintering furnace. Some simplifying hypotheses are made in the furnace modelling for an initial steady state study:

- Entrance and exit doors are closed;
- The walking beam mechanism is not modelled, and the beam is always considered in the upper position. The lower part of the walking beam and all the displacement mechanism are not represented;
- All the joints between the refractory bricks are not represented. In cold conditions, their thickness is below 2 mm and at nominal operation, with the dilatation of bricks, it becomes even thinner. Therefore, one can assume their influence on the thermal analysis is negligible;
- All the pipes bringing the gas through the bricks for injection or thermocouples measuring the temperature in the pre-sintering and sintering zones are not represented. Their explicit modelling is too consuming in mesh and computing time, and their influence is negligible, which makes their representation not worth the effort;
- The outside casing and the insulating layer under the casing are not represented. However, their thermal resistance and emissivity will be considered in the boundary limits of the furnace;
- The heating elements were represented with a simplified model at first, but because of important thermal radiation effects, a more realistic shape was needed (see Section 4.2).

The computational domain displayed is composed of two parts:

- The solid domain, displayed in beige in Figure 2, is made of bricks. Several types of bricks are used: the ones near the core differ from those placed on the edge of the furnace, with densities decreasing between the core to the edge. Each brick's specific thermal properties are considered in the modelling;
- The gas domain displayed in blue in Figure 2, extends around the furnace refractory bricks through open areas in front, behind, and below, respectively. This is due to the entrance zone, the cooling zone, and the volume containing the displacement mechanism of the walking beam. These parts are only limited by the casing not explicitly represented as mentioned above.

The grid is composed of polyhedric cells. In the solid domain, at least three layers of cells are ensured to resolve properly the heat equation. In the gas, the grid is finer to be more precise. A low Reynolds number indicates that the flow is rather laminar in the furnace. The grid contains 11.2 million mesh cells.

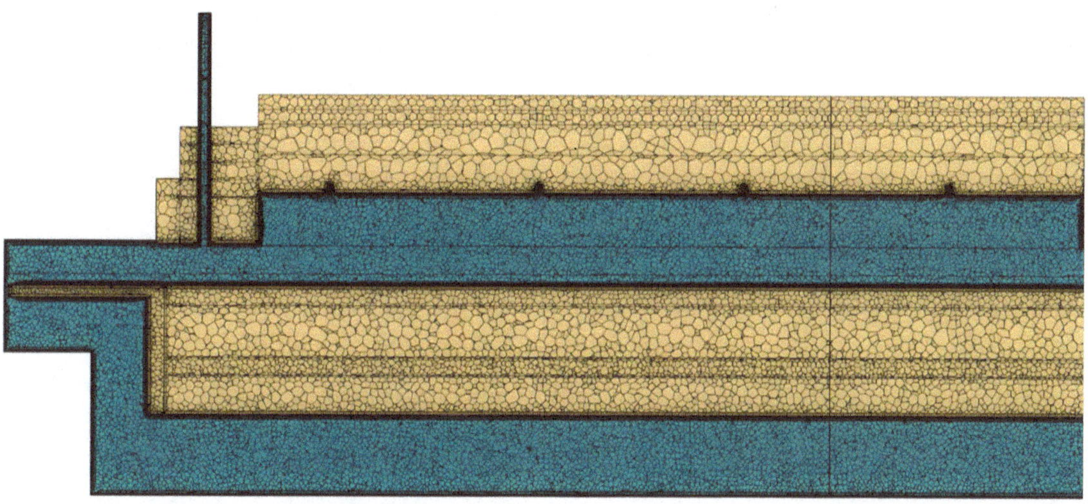

FIG. 2. Example of a longitudinal section of the grid (in blue: gas, in beige solid).

4.1 Boats with equivalent UO_2 product

In the first approach, the furnace was considered empty. Therefore, boats and pellets were not represented. Then boats with equivalent UO_2 products are included (Figure 3). The exact representation of each pellet is not yet modelled for computational cost reasons. It is possible to model:

- A sintering furnace full of boats or also choose a partially empty furnace;
- Boats with different equivalent filling ratio to represent the quantity of pellets inside a boat.

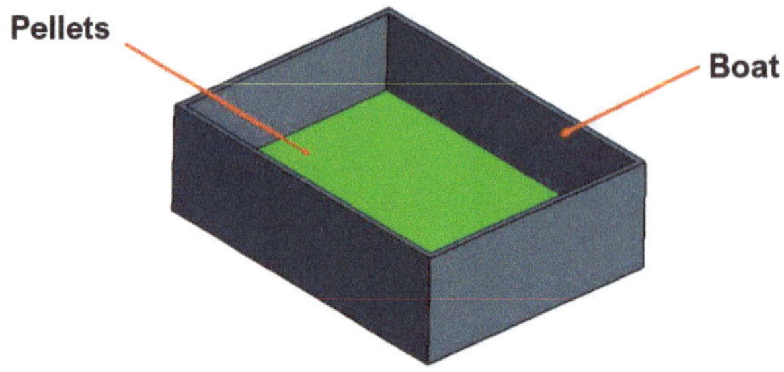

FIG. 3. Representation of the modelled boat and its pellets.

These boats and their pellets are added for two reasons:

- To have a more realistic gas flow in the furnace. Indeed, the boats significantly reduce the flow area in some sections of the furnace for instance in the vestibule at the transition zone between the pre-sintering and the sintering zones;
- To consider a part of their thermal effects on the total heat balance. A part of the furnace power is used to heat the pellets.

4.2 Heating elements

In the first Computational Fluid Dynamics model, the heating elements are modelled by solid blocks. Because of the involved temperatures in the domain, radiation transfer is the dominant heat transfer mode. The initial model with blocks was designed to facilitate the conduction of the heating elements with the adjacent bricks through the lateral faces of the blocks. However, this approach was not physically accurate and resulted in a reduction in the radiation heat transfer. Furthermore, as the shape of the heating elements matters for radiation transfer, the blocks have been replaced by a more realistic shape. The results of the initial model with blocks and the new one with more realistic heating elements are shown in Figure 4. The model is more accurate and closer to the measured temperatures, especially in the pre-sintering area. By improving the modelling, another advantage is the better representation of the flow as the gases can now propagate in the heating elements sockets.

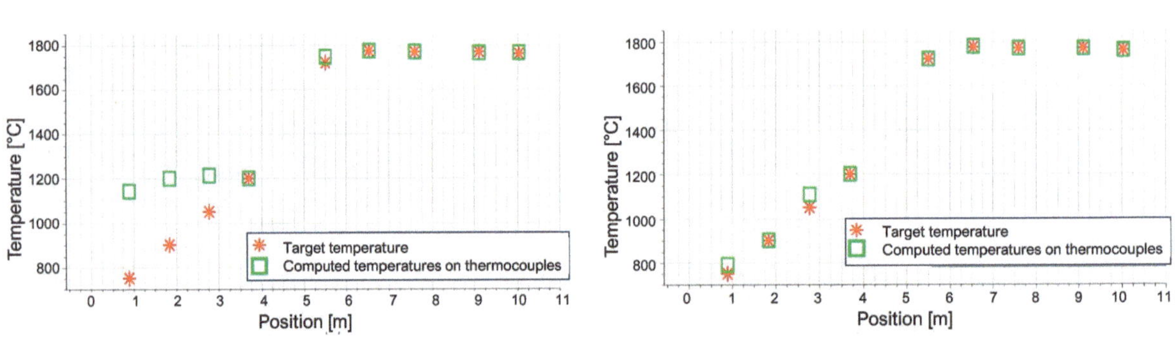

a) First modelling with a simple block b) Realistic shape modelling of the heating elements

FIG. 4. Comparison of the computed temperatures with the targeted ones on the thermocouples for the 2023 and 2022 models.

4.3 Results

An example of the global thermal fields is given in Figure 5. With the current model available:

- The thermal flow is rather uniform in the sintering zone. However, in the pre-sintering zone, heterogeneities of temperature appear clearly. The hot gas coming from the sintering zone and ejected through the pre-sintering chimney disrupts the flow and creates important gradients;
- The targeted temperature profile is computed on the thermocouples, represented by red dots in Figure 5, along the furnace. From the observation that by regulating the heating elements powers on the temperature computed on the thermocouples, the boats of pellets are cooler than expected due to the important gas stratification developing in the furnace vein. A new calculation is made by regulating the powers directly with the average temperature of the boats located under the thermocouples. This hypothesis allows us to match perfectly with the power profile of the operational data.

The heating element powers regulated by the thermocouples correspond to the correct orders of magnitude.

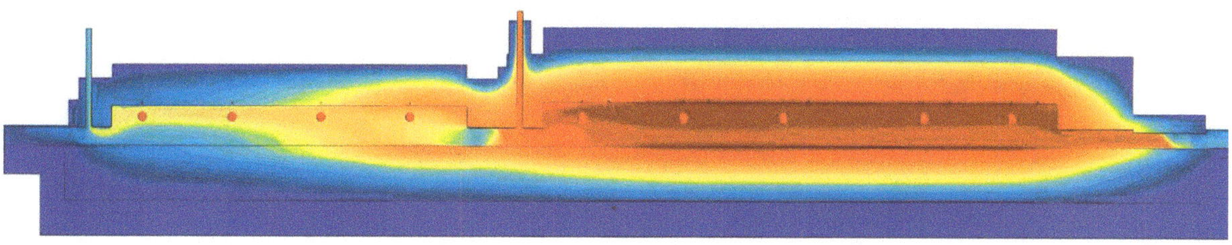

FIG. 5. Temperature fields on the longitudinal section (gas and solid). The colour blue is used to represent low values, while the colour red is used to represent high values.

5. CONCLUSION AND OUTLOOKS

The sintering process is complex with sublimation and reduction reactions involved, but also densification and grain growth. With the current model available it is possible to determine the temperatures and gas flows at the different locations in the furnace and to perform sensitivity studies. This could lead improving the sinter furnace use and its performance by optimizing the process parameters to enhance pellet quality, by increasing the throughput, by limiting the electrical need, thus production costs, and by reducing the furnace aging.

There are many benefits to modelling the key steps of the process such as:

- Perform parametric studies without using the production equipment.
- Increase performance by optimizing the process, e.g. decrease the quantity of energy needed to run the sintering furnaces;
- Compare the different equipment available among Framatome and use the best developments;
- Create a teaching tool through the theoretical description of the process and its key influencing parameters.

The outlooks for this study could be:

- Introduce chemical reactions to reproduce sublimation, reduction reactions, and microstructure changes;
- Create a multiscale model to collect information at boat scale, pellet scale, grain scale, and atomic scale;
- Add different products (Gadolinium pellets, Cr_2O_3-doped UO_2 pellets) and adapt the blend recipe (additives, U_3O_8 content);
- Model other equipment considered as a key step in the pellet production process such as compaction with the press with experimental data in parallel.

This R&D development is part of Framatome's dedication to continuous improvement to increase performance and deliver quality products to its customers.

REFERENCES

[1] INTERNATIONAL ATOMIC ENERGY AGENCY, Experiences and Trends of Manufacturing Technology of Advanced Nuclear Fuels, IAEA-TECDOC-1686, IAEA, Vienna (2012).

[2] OLEVSKY, E. A., Theory of sintering: from discrete to continuum, Mater. Sci. Eng.: R: Reports, **23** 2 (2018) 41-100.

[3] GREENQUIST, I., et al.; Review of sintering and densification in nuclear fuels: Physical mechanisms, experimental results, and computational models, J. Nucl. Mater. **507** (2018) 381-395.

ID#6

EVALUATION OF THE CORRELATION BETWEEN GREEN AND SINTERED DENSITY IN UO2 PELLETS

J. P. R. CARNAVAL
Industrias Nucleares do Brasil (INB),
Resende, RJ, Brazil

Abstract

The production of UO_2 fuel pellets is a critical process in nuclear fuel manufacturing, impacting reactor performance and safety. This study focuses on understanding and controlling the factors influencing the density of sintered pellets, crucial for their performance in nuclear reactors. Operational tests were conducted using various mixtures of UO_2, U_3O_8, and dopants, with the density of green pellets (GD) as a key parameter.

Results indicate a linear correlation between sintered density (SD) and green density (GD) for UO_2 pellets, allowing for the establishment of equations to predict SD based on GD. Additionally, the impact of U_3O_8 content and dopants on SD was investigated. It was found that higher U_3O_8 content and certain dopants decrease SD, both exhibiting a stronger influence than GD. For Angra 1 and Angra 2 pellet designs, optimal GD ranges were identified to achieve target SD values. Furthermore, dopants were shown to significantly affect grain size playing an important role in increasing product quality.

In conclusion, GD is one of the main parameters to be adjusted in a pressing process to control the SD of UO_2 pellets. U_3O_8 and dopants have a greater impact on sintered pellets, but GD still serves as a valuable parameter for controlling SD, offering additional adjustments. Understanding these relationships enhances the production process, ensuring the reliability and safety of nuclear reactors. This research provides valuable insights for optimizing fuel pellet manufacturing and reactor performance.

1. INTRODUCTION

The production of fuel elements at INB includes the manufacturing stage of UO_2 pellets, which are the basis for energy production at the Angra NPPs. This component is specified in accordance with the relevant standards [1] which density is one of the main characteristics of the product. To achieve the specified range, the process steps need to respect qualified parameters and allow adjustments within operational control ranges when pressing the master mixture UO_2-U_3O_8-ADS (Aluminium Distearate). Pressing is a process in which an amount of powder is fed into the hole of the die where the lower and upper punches slide to compact and form the pressed pellet commonly called green pellet. This compact is an agglomeration of powder particles with little mechanical resistance and can fall apart when handled. Each compacted piece can have an independent green density, normally varying within an established range so that, after sintering, it acquires permanent ceramic properties and reaches the final density suitable for use in fuel elements. Therefore, the sintered density range strongly depends on the green density range resulting from the press acting on the inserted powder.

One of the main techniques for pressing powders into shapes is the isostatic pressing which applies pressure uniformly in all directions, minimizing density gradients and certain defects associated with uniaxial pressing. However, even isostatically pressed compacts can exhibit density gradients due to interparticle frictional forces.

The main defects resulting from the uniaxial pressing process are studied to better understand the causes: Cracks often originate from poor interparticle bonding during compaction, while warping can result from factors like green density gradient, friction drag, gravity, and temperature gradient during sintering. Friction between powder and die wall is identified as a significant factor influencing compaction uniformity and density gradients in pressed compacts. This friction can be minimized with smooth surface dies, carbide tooling, and the use of lubricants. However, the complete elimination of frictional effects is very difficult to reach.

The impact of these defects is particularly significant in the context of pellet cladding interaction margin reduction. Pellet cladding interaction refers to the interaction between the fuel pellet and the cladding (outer layer) of the fuel rods in a nuclear reactor. If there are defects in the fuel pellets, such as cracks or irregularities, they can exacerbate the interaction with the cladding during reactor operation.

When the fuel pellets interact with the cladding, they can cause mechanical stresses and thermal changes in the cladding material. If these stresses exceed certain limits, it can lead to the degradation of the cladding material, potentially resulting in breaches or failures. This can compromise the integrity of the fuel rods and pose safety risks in the nuclear reactor. Hence, it's important to minimize fabrication-induced defects in ceramic fuel pellets to ensure the reliability and safety of nuclear reactor operation.

The resultant defects on UO_2 pellets may be significantly reduced by acting on the main sources of variation: Process, Material and Operation. The pressing process can be adjusted to reduce the final number of defects by optimizing the pressing parameters. On the other hand, the raw material can contribute to the quality of the parts, and the characteristics of UO_2, the quantity of U_3O_8 and the possibility of using dopants have to be considered.

At INB, the green pellets are submitted to the sintering process at 1750°C and 4 hours (higher heat zone) to obtain ceramic properties. At this temperature, the UO_2 diffusion causes the grain growth and the pores are consumed along this process. The U_3O_8 is completely converted to UO_2 and residual oxygen, also pores are formed. This UO_2 is easily incorporated into the UO_2 matrix which continues the sintering process while the pores are usually consumed by the densification process.

The fuel pellet sintering process normally uses parameters that generate UO_2 pellets with adequate density for nuclear reactors depending on the desired working power. Additionally, the pores are also capable of accumulating a certain amount of fission gases released during the burnup of fuel in the reactor. As a result, the presence of pores is desired and thus the use of U_3O_8 in the production of pellets is also a topic studied in this article.

The use of dopants in nuclear UO_2 pellets can indeed help reduce cracking phenomena during the manufacturing and operational stages. Dopants, also known as additives, are introduced into the uranium dioxide (UO_2) matrix to alter its properties. The effect of dopants depends on the type and quantity of dopant used. They can increase UO_2 average grain size and modify pore formation mechanisms during sintering. This results in a more uniform microstructure with fewer voids and defects, which can help mitigate the occurrence of cracks. Moreover, certain dopants can improve the thermal conductivity of UO_2 pellets. This helps in more uniform heat distribution within the pellet during operation, minimizing temperature gradients that can contribute to cracking. Thus, dopants may alter the mechanical properties of UO_2, including its elastic modulus and thermal expansion coefficient. By modifying these properties, dopants can help the pellet accommodate thermal and mechanical stresses more effectively, reducing the likelihood of crack initiation and propagation.

This work presents the results of operational tests with the manufacture of fuel pellets at INB that allow the evaluation of the correlation between green density and sintered density in UO_2 pellets considering the following points:

- Variations in Density of Green Pellets – Process parameters;
- Variation in composition of UO_2 and U3O8 mixtures – Materials;
- Addition of dopants to mixtures – Process and Materials.

The overall objective is to understand how sintered density is affected by green density and obtain the best operational range. Furthermore, another objective is to understand how mixture variations can be used to overcome the problems caused by natural green density variations in the fuel pellet manufacturing process.

2. DEVELOPMENT AND TESTS

A sequence of operational tests was developed with UO_2, U_3O_8 and dopants on this work. Different combinations of powder mixtures were proposed, and all of them were pressed in a Courtoy R53 model rotary press, in which the parameters were configured for a green pellet density range.

2.1. Materials

Mix I: UO_2 Powder (99.8%) and ADS (0.2%).

Mix II: UO_2 and U_3O_8:

A. UO_2 Powder (87.8%), ADS (0.2%) and 12% U_3O_8;
B. UO_2 Powder (89.8%), ADS (0.2%) and 10% U_3O_8.

Mix III: UO_2 Powder (~ 99.5%), ADS (0.2%) and dopants:

A. + Al_2O_3 – 0.05%, 0.1%, 0.2%, 0.3%;
B. + Nb_2O_5 – 0.05%, 0.1%, 0.2%, 0.3%;
C. + Cr_2O_3 – 0.05%, 0.1%, 0.2%, 0.3%.

Mix IV: UO_2 Powder (~87.5%), ADS (0.2%), 12% U_3O_8 and dopants:

A. + Al_2O_3 – 0.1%, 0.2%, 0.3%, 0.5%;

B. + Nb_2O_5 – 0.1%, 0.2%, 0.3%, 0.5%;

C. + Cr_2O_3 – 0.1%, 0.2%, 0.3%, 0.5%.

2.2. Process Parameters

- Density of Green Pellets: 5.6–5.9 g/cm^3 (51–54% TD);
- Sintering: Temperature at top heat zone in the Furnace: 1760°C; 5 hours.

3. RESULTS AND DISCUSSION

3.1. Evaluation of Density on Sintered Pellets produced with mixtures I and II.

3.1.1. Correlation between Green and Sintered Density and U_3O_8 Effects

The results of the density analysis on sintered pellets as a function of the density of the green pellets (Mix I) are presented in Figure 1, as well as the data for pellets produced with mixes II.A and II.B, with 10% and 12% U_3O_8.

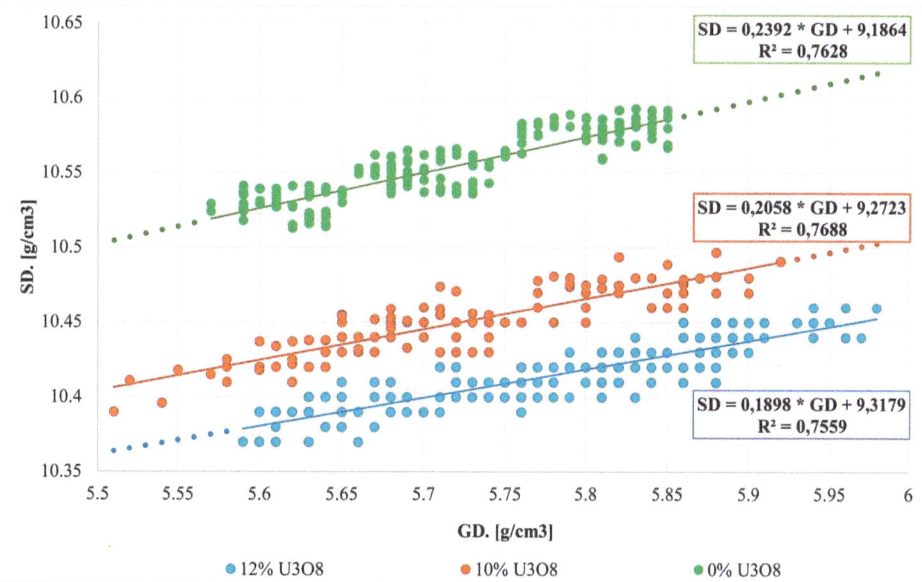

FIG. 1. Correlation between Green (GD) and Sintered Density (SD) for 3 levels of U_3O_8.

As demonstrated in Figure 1, there is a linear correlation between sintered density (hereby, SD) and green density (hereby, GD) for UO_2 pellets made from UO_2 or UO_2–U_3O_8 blends. This data leads to the establishment of a sintered density equation as a function of green density for each type of mixture. The equations allow using green density as an input parameter for the production of sintered pellets in a range suitable for each type of fuel that is desired to be produced. Furthermore, it is evident that an increase in the quantity of U_3O_8 in the powder mixture tends to decrease the density of the sintered pellet, which is expected since U_3O_8 is pore-forming.

The pellets manufactured at INB aim to meet two projects, with nominal sintered densities at 10.40 and 10.45 g/cm^3. In order to achieve these parameters, the density data were isolated in graphs according to the U_3O_8 content in the mixtures as shown in the following sections.

3.1.2. Equation for SDxGD – Angra 1 fuel design

Figure 2 presents the sintered density equation as a function of the green density for pellets with an SD target at 10.40 g/cm^3, typically used for Angra 1 fuel design. The use of 12% U_3O_8 is recommended, establishing the optimal working range for the green density (in blue) that leads to the desired sintered density. It is also possible to understand that the same mixture can be used in the production of pellets with a sintered density in the order of 10.45 g/cm^3 by increasing the green density (in green), confirming that the green density can be directly used as a parameter for controlling the production of fuel pellets.

As a brief conclusion, for Angra 1 pellet design (nominal SD = 10.40 g/cm^3) the GD optimum range is 5.65–5.75 g/cm^3 with a material composition corresponding to 88% UO_2: 12%U_3O_8.

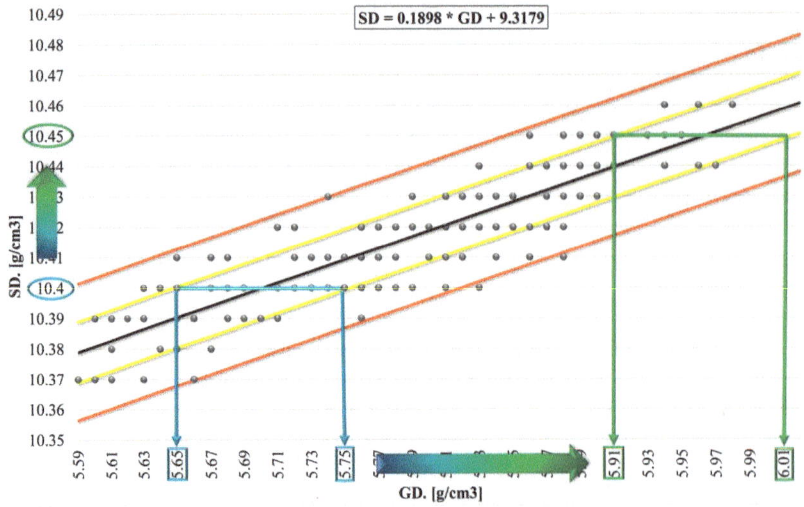

FIG. 2. Correlation SDxGD – Angra 1 fuel design – target SD = 10.40 g/cm^3 – 12% U_3O_8 (Mix II.A).

3.1.3. Equation for SDxGD – Angra 2 fuel design

Figure 3 presents the sintered density equation as a function of the green density for pellets with an SD target of 10.45 g/cm^3, typically used for Angra 2 fuel design. The use of 10% U_3O_8 is recommended, establishing the optimal working range for the green density (in blue) that leads to the desired sintered density. It is also possible to understand that the same mixture could be used in the production of pellets with sintered density in the order of 10.40 g/cm^3 by decreasing the green density (in purple), confirming that the green density can be directly used as a parameter for controlling the production of fuel pellets.

As a brief conclusion, for Angra 2 pellet design (nominal SD = 10.45 g/cm^3) the GD optimum range is 5.67–5.77 g/cm^3 with material composition corresponding to 90% UO_2: 10% U_3O_8.

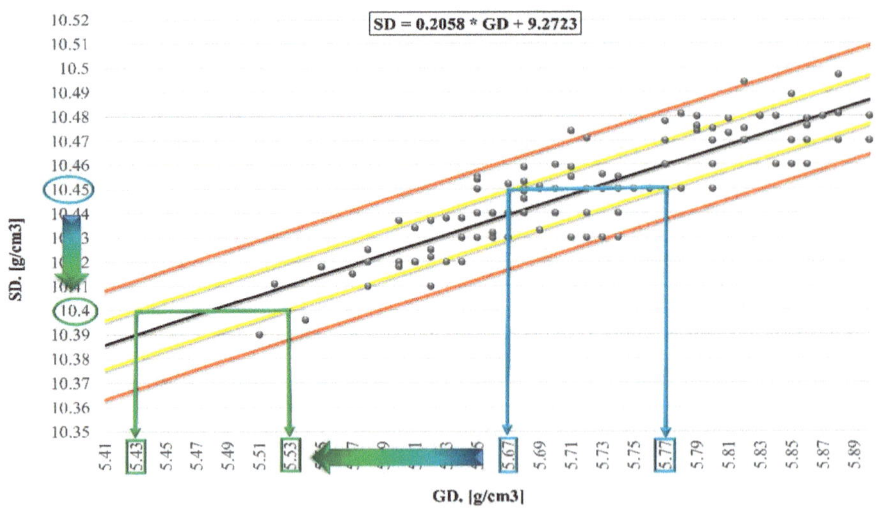

FIG. 3. Equation for SDxGD – Angra 2 fuel design – target SD= 10.45 g/cm^3 – 10% U_3O_8 (Mix II.B).

3.1.4. Evaluation of the Main Effects and Interaction Plot for SD as a function of GD and U_3O_8 content

The results for SD measured on pellets manufactured with mixtures I and II (same data used at items 3.1 to 3.3) were compiled in Figure 4. The SD data were organized by fitted means as a function of GD and U_3O_8 content in such a way that it is possible to clearly observe the 3 main items:

✓ On the left, how the increase in U_3O_8 content implies a reduction in sintered density;

✓ In the centre, how the increase in GD implies an increase in SD for any concentration of U_3O_8;

✓ On the right, the combination of the two: The increase in SD as a function of GD (coloured curves) for each U_3O_8 content.

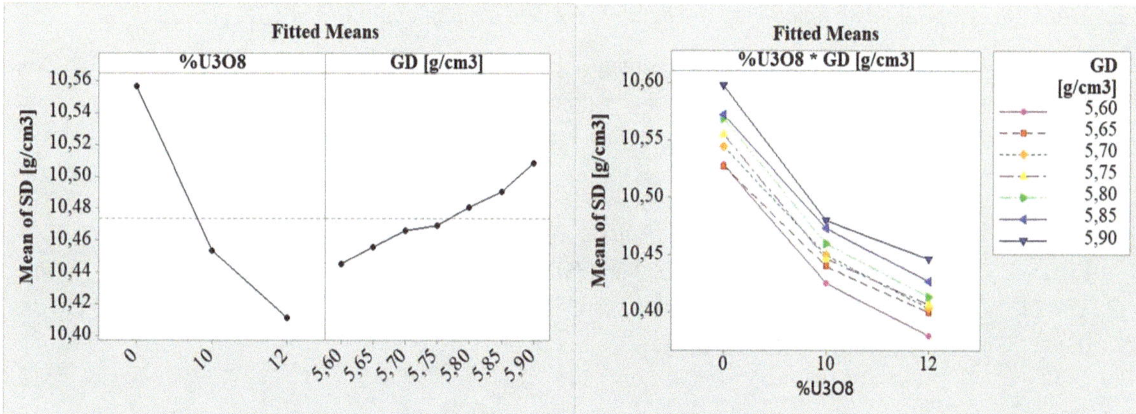

FIG. 4. Fitted Means for SD as a function of GD and U_3O_8 content - Main Effects (left) and Interaction Plot (right).

The impact of adding up to 12% U_3O_8 on pellets is approximately double that caused by the GD range (5.6–5.9 g/cm^3) and the U_3O_8 content can be used to control SD of pellets. As a result, setting appropriate GD range works as fine adjustment to obtain the target value of sintered pellets.

3.2. Evaluation of Density on Sintered Pellets produced with mixture III.

The results of density analysis on sintered pellets as a function of dopant content, type of dopants and GD were used to generate Figure 5. The pellets were manufactured from mixtures III (A, B and C) with UO_2 (above 99.5%), 0.2% ADS and dopants from 0 to 0.3% weight, all of them 0% U_3O_8.

3.2.1. Main Effects and Interaction Plot for SD as a function of the dopants content, type of dopants and GD

The evaluation of the effects demonstrated by Figure 5 '*left*' allows us to understand the behaviour of SD as a function of GD in doped pellets. Additionally, the dopants utilized in this study (up to 0.3% Al, Nb, Cr) have been demonstrated to achieve an SD increase that is twice that of GD (range 5.6–5.9 g/cm^3). Even thus, it is possible to observe the existence of a linear correlation between SD and GD and this variable may also be used to control SD of doped pellets as a fine adjustment.

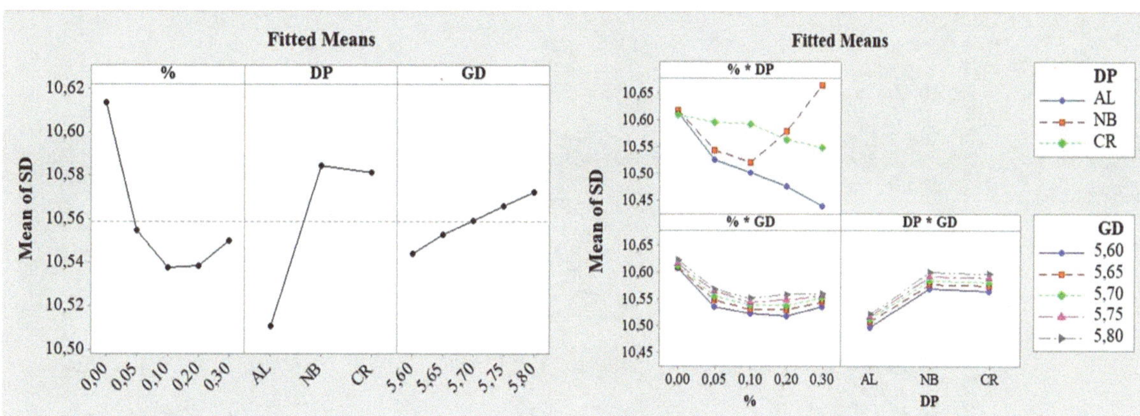

FIG. 5. Fitted Means for SD as a function of dopant content, dopant type, and GD range - All 0%U_3O_8 Main Effects (left) and Interaction Plot (right).

On the right side of Figure 5, it is possible to observe the effect of each type of dopant and their respective quantities, as well as their interactions, on the sintered density: increasing the Al and Cr content tends to reduce SD, but increasing Nb content can reduce and increase SD, a phenomenon that is still being studied. Once more, it is evident that the impact of the dopant on the pellet is greater than that of GD. Consequently, GD can be employed as a suitable parameter for SD adjustment on doped pellets.

3.2.2. Main Effects for SD as a function of the dopants content, type of dopants and GD

The results for SD measured on pellets manufactured with mixture III (same data used in section 3.2.1) were compiled in Figures 6, 7 and 8. The SD data were organized by fitted means as a function of the dopants content, type of dopants and GD in such a way that it is possible to clearly observe the 3 main facts:

- ✓ Figure 6: The increase in Al, Cr, Nb content implies a reduction in sintered density;
- ✓ Figure 7: As the type of the dopant leads to an increase in SD in this sequence: Cr>Nb>Al;
- ✓ Figure 8: The linear correlation between SD and GD;
- ✓ The SD of pellets with 12% U_3O_8 are considerably lower than that of pellets with 0% U_3O_8, regardless of dopant content, dopant type and GD range.

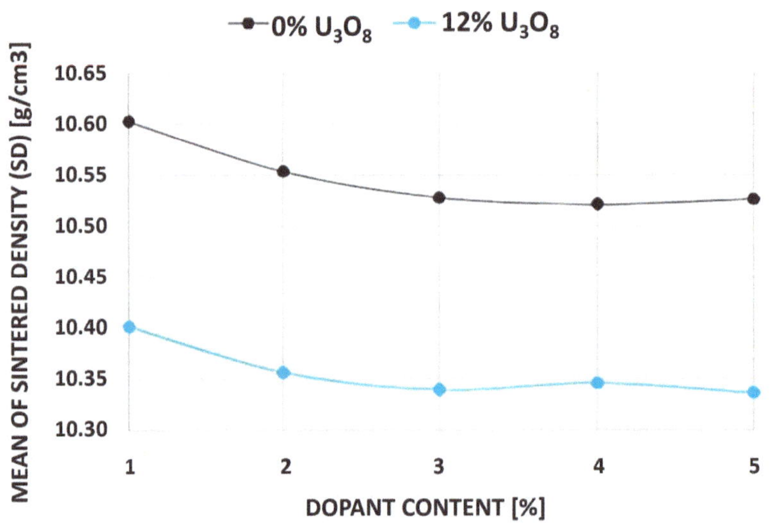

FIG. 6. Fitted Means for *SD as a function of dopant content for mixtures with 0%U_3O_8 and 12%U_3O_8.*

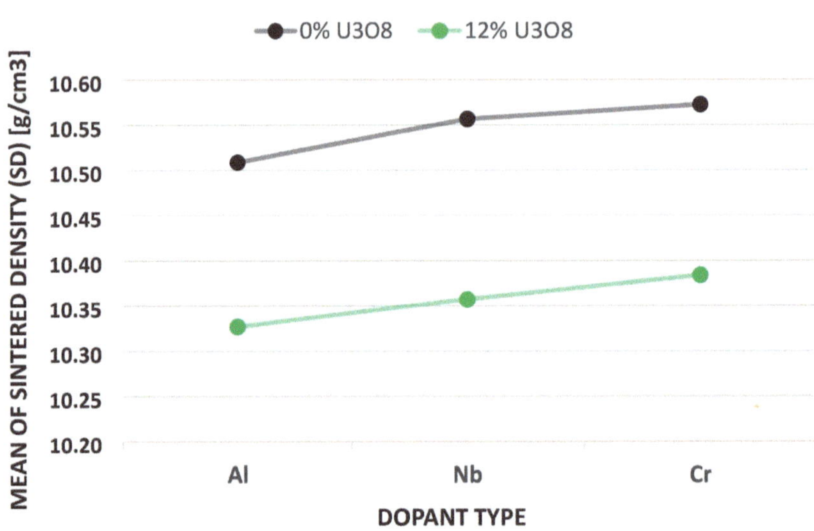

FIG. 7. Fitted Means for *SD as a function of dopant type for mixtures with 0%U_3O_8 and 12%U_3O_8.*

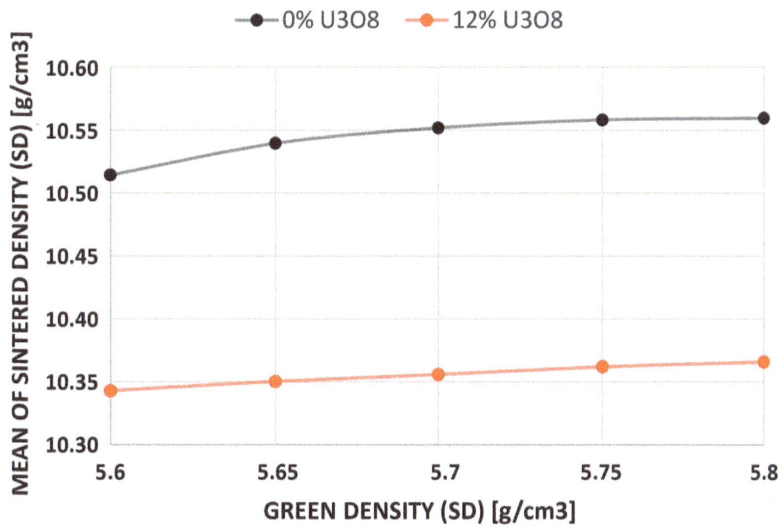

FIG. 8. Fitted Means for SD as a function of GD range for mixtures with 0%U_3O_8 and 12%U_3O_8.

Green density is always found as a parameter that linearly affects SD, but it contributes only slightly when compared to the impact caused by the use of U_3O_8 and/or dopants, However, it can be utilized to accommodate the SD of doped pellets as a process adjustment.

3.3. Evaluation of Sintered Density and Grain Size on Pellets produced with mixture IV.

The results of sintered density and grain size analysis on sintered pellets as a function of dopant content and type were used to generate Figures 9, 10 and 11. The pellets were manufactured from mixtures III (A, B and C) with UO_2, ADS and dopants; and mixtures IV (A, B and C) with UO_2, ADS, 12% U_3O_8 and dopants from 0 to 0.5% weight. The GD range was adjusted to the optimum range (5.65–5.75 g/cm^3).

3.3.1. Sintered Density (SD) and Grain Size (GS) of Pellets Manufactured with Al_2O_3

The curves for SD (black) and GS (blue) of pellets doped with Al_2O_3 are plotted on Figure 9. The increase in Al_2O_3 content leads to a decrease of SD at both U_3O_8 quantities and causes a variation in the grain size mainly when no U_3O_8 is used. On the other hand, the GS looks stabilized when using 12% U_3O_8. Both SD and GS are lower when using 12% U_3O_8 for any Al_2O_3 content. The green density does not seem to have a significant effect on GS but yet may be used to fit SD of doped pellets as a process adjustment.

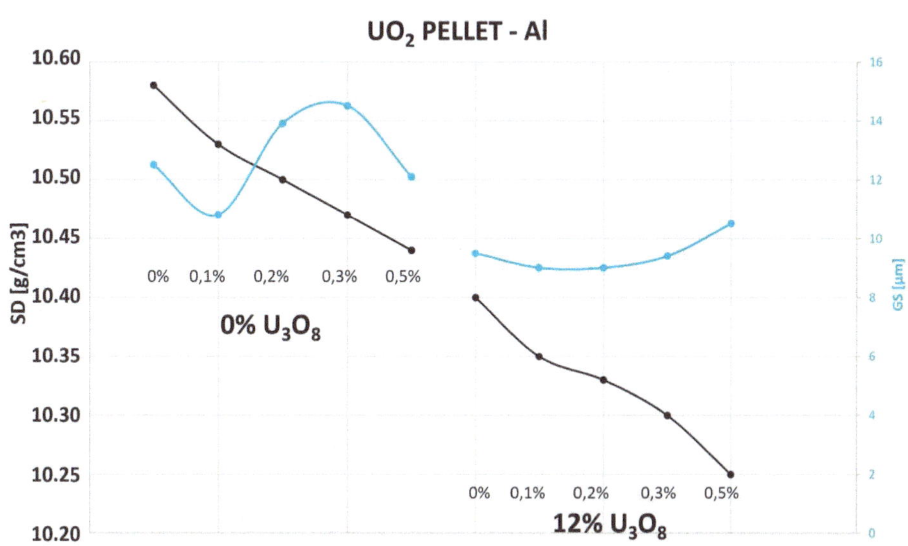

FIG. 9. Sintered Density (left) and Mean Grain Size (right) as a function of Al_2O_3 content and U_3O_8 content.

3.3.2. Sintered Density (SD) and Grain Size (GS) of Pellets Manufactured with Nb_2O_5

The curves for SD (black) and GS (green) of pellets doped with Nb_2O_3 are plotted in Figures 10. The increase in Nb_2O_5 content causes a variation in SD at both U_3O_8 quantities and increases almost linearly the grain size. The SD variation caused by Nb_2O_5 is being studied but it would be feasible to admit that the increase in the content of this dopant can influence the formation of pores in the matrix, causing a reduction in density, even with the average increase in grain size. Both SD and GS tend to decrease value when using 12% U_3O_8. Once again, the green density does not seem to have a significant effect on GS yet may be used to fit SD of dopped pellets as a process adjustment.

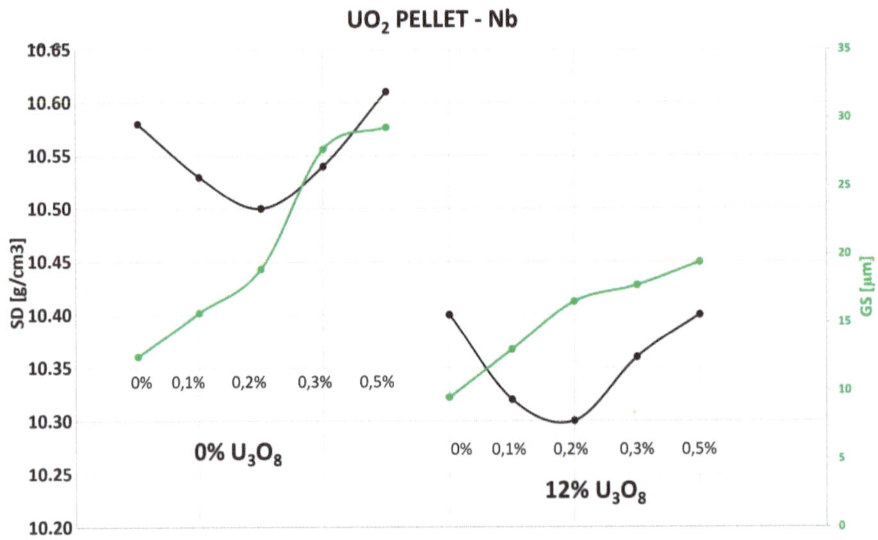

FIG. 10. Sintered Density (left) and Mean Grain Size (right) as a function of Nb_2O_5 content and %U_3O_8.

3.3.3. Sintered Density (SD) and Grain Size (GS) of Pellets Manufactured with Cr_2O_3

The curves for SD (black) and GS (red) of pellets dopped with Cr_2O_3 are plotted in Figures 11. The increase in Cr_2O_3 content decreases the SD at both U_3O_8 quantities and increases almost linearly the grain size. The smooth SD variation caused by Cr_2O_3 when using no U_3O_8 may be also related to a slight pore formation. Both SD and GS tend to decrease value when using 12% U_3O_8. No pellets with 0.5% Cr_2O_3 were produced, which is why the yellow dot is just an estimate based on the data obtained. Once again, the green density does not seem to have a significant effect on GS yet may be used to fit SD of dopped pellets as a process adjustment.

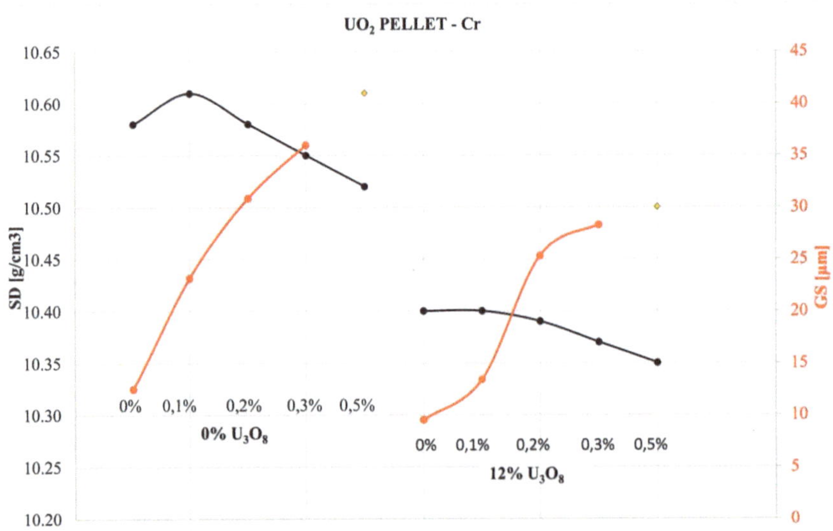

FIG. 11. Sintered Density (left) and Mean Grain Size (right) as a function of Cr_2O_3 content and %U_3O_8.

4. CONCLUSIONS

There is a linear correlation between density of sintered pellets and green pellets made from UO_2 or UO_2–U_3O_8 blends.

This data leads to the establishment of a sintered density equation as a function of green density for each type of mixture. The equations allow the use of green density as an input parameter to produce sintered pellets in a range suitable for each type of fuel that is desired to be produced. At INB, the following conclusions were found:

For Angra 1 pellet design (nominal SD = 10.40 g/cm^3) the GD optimum range is 5,65–5,75 g/cm^3 with material proportion 88% UO_2: 12%U_3O_8. For Angra 2 pellet design (nominal SD = 10.45 g/cm^3) the GD optimum range is 5,67–5,77 g/cm^3 with material proportion 90% UO_2: 10%U_3O_8.

The use of up to 12% U_3O_8 on pellets can achieve twice the SD increase compared to GD (range 5.6–5.9 g/cm^3) and the U_3O_8 content can be used to control SD of pellets.

The dopants (up to 0.5% Al_2O_3, Nb_2O_5, up to 0.3% Cr_2O_3) can achieve twice the SD increase compared to GD (range 5.6–5.9 g/ cm^3) and dopants can be used to control SD of pellets.

The SD of pellets with 12% U_3O_8 are considerably lower than that of pellets with 0% U_3O_8, regardless of dopant content, dopant type and GD range. The increase of Nb_2O_5 and Cr_2O_3 content leads to an increase in the grain size causing some variations of SD (decreasing it in some cases) but it´s considered beneficial to the pellet since a larger grain size generates a safer product as it can influence the fission gas release. The variation in SD caused by dopants could be softened by the U_3O_8 content.

Finally, all tests reported the existence of linear correlations between SD and GD which could be always used to control SD of doped pellets at any U_3O_8 content as a fine adjustment to obtain the target value.

REFERENCES

[1] AMERICAN SOCIETY FOR TESTING AND MATERIALS, Standard Specification for Sintered Uranium Dioxide Pellets for Light Water Reactors, ASTM C776, ASTM International

[2] INB, INDÚSTRIAS NUCLEARES DO BRASIL, Ciclo do combustível nuclear, (June 2023) http://www.inb.gov.br/pt-br/Nossas-Atividades/Ciclo-do-combustivel-nuclear.

[3] OCCHIONERO, M. A., HALLORAN J.W., The Influence of Green Density Upon Sintering, Case Western Reserve University, Cleveland, Ohio **16** (1984) 89-121.

[4] COSTA, D. R., Manufacturing of doped UO_2 fuel ceramic pellets for nuclear application, COPPE, UFRJ. Rio de Janeiro (2018).

[5] TURNBULL, J. A., The Effect of Grain Size on the Swelling and Gas Release Properties of UO_2 During Irradiation, J. Nucl. Mater. **50** (1974) 62-68.

[6] SONG, K. W., et al., Sintering of Mixed UO_2 and U_3O_8 Powder Compacts, J. Nucl. Mater. **277** (2000) 23-129.

[7] OHÂI, D., Large Grain Size UO_2 Sintered Pellets Obtaining Used for Burn up Extension, Transactions of the 17th International Conference on Structural Mechanics in Reactor Technology (SMiRT 17), Prague, Czech Republic, August 17-22, Paper # C02-3 (2003).

ID#7

ADVANCED FABRICATION AND CHARACTERIZATION OF GD PELLETS FOR SMR FUEL IN ARGENTINA

M. F. PARRADO
Advanced Fuels Department,
Comisión Nacional de Energía Atómica (CNEA)
Buenos Aires, Argentina

Abstract

The paper describes the different stages of research, development, fabrication and characterization of Gd pellets for an SMR fuel in Argentina. The development of uranium dioxide fuel pellets with burnable poison covered a first laboratory scale phase, a change of scale in the process, and an adaptation of facilities and equipment in view of their manufacture to supply the CAREM 25 prototype reactor. Also, for the provision of fuel pellets for the CAREM 25 reactor, the Department carried out a survey and update on the characterization techniques involved in the process and implemented a quality system relevant to the project's requirements.

1. INTRODUCTION

The National Atomic Energy Commission (CNEA, by its Spanish acronym) has among its objectives the research and development of the peaceful uses of nuclear energy in our country, promoting innovation through basic and applied science in different scientific fields and the productive structure.

The CNEA carries out multiple projects on the application of nuclear technology. Moreover, it has worked on the design, construction, and operation of research reactors and radioisotope production; the manufacture of nuclear fuels, heavy water, and cobalt 60; and the construction of solar panels for Argentine satellites, among other scientific and technological developments.

Currently, Argentina has three NPPs: Embalse, Atucha 1, and Atucha 2. In the field of SMRs, the development of the CAREM reactor is in progress and it will be the first nuclear power reactor entirely designed and built in Argentina. With this new milestone, our country consolidates its capacity for the development and commissioning of NPPs, and a key role in the use of nuclear energy for electricity production.

The Advanced Fuels Department, belonging to the CNEA Nuclear Fuel Cycle Area Management, developed uranium dioxide fuel pellets with burnable neutron poisons for an SMR reactor, and defined a methodology and manufacturing parameters compatible with technologies already installed in local industry.

2. FUEL PELLETS DEVELOPMENT FOR A SMR

The Advanced Fuels Department was in charge of the development of fuel pellets for a SMR reactor, defining the characteristics, methodology and manufacturing parameters compatible with technologies already installed in local industry and controls to be carried out to obtain them uniformly and on an industrial scale. The SMR fuel pellets involve two lines of development: enriched uranium dioxide pellets and uranium dioxide–gadolinium oxide pellets (UO_2–Gd_2O_3). On this paper we will focus on the uranium dioxide– gadolinium oxide pellets (UO_2–Gd_2O_3) development, and only mention a brief description of the enriched uranium dioxide pellets development steps.

2.1 Enriched UO_2 fuel pellets for a SMR

2.1.1 Fuel pellets development

It consists of the development of 3,1% enriched UO_2 fuel pellets with pore former. The development includes the characterization of the UO_2 powder, the study and selection of pore formers and lubricants, the mechanical mixtures tests and the study of pressing and sintering parameters.

2.1.2 Development and validation of characterization techniques for powders and pellets

The development of characterization techniques includes the characterization of raw materials powders, intermediate products and fuel pellets in all stages of the process, according to the technical specifications. In addition, we conducted a survey of CNEA laboratories qualified to perform the tests.

2.1.3 Fuel pellets manufacture for irradiation in HALDEN reactor

As a milestone to highlight within the development stage, 200 3.1% UO_2 fuel pellets were manufactured to be irradiated in the Halden reactor (Norway).

2.1.4 Pilot scale development

Once the development stage was completed, the change of scale stage began, which included the mixing system change of scale, dies and press automatic loading system adjustment, the definition of manufacturing control parameters, the definition of continuous sintering parameters and grinder adjustment. On the pilot scale development, 200 kg of material was processed, obtaining satisfactory results consistent with those obtained in the development stage. In this way, a manufacturing technology was defined.

2.1.5 Technology transfer to CONUAR

Once the manufacturing technology of manufacturing for producing 3.1% enriched UO_2 fuel pellets with pore former was established, the process of transferring the technology to the factory began. It included technical meetings and assistance to CONUAR in preparing the manufacturing line. Also, it involved the preparation and delivery of documentation (manufacturing procedures, pilot scale manufacturing results reports and fuel pellets characterization).

2.2 UO_2 – Gd_2O_3 fuel pellets for a SMR

This line of development involves natural UO_2 pellets with Gd_2O_3 as burnable poison. These pellets were developed in the Burnable Poisons Laboratory, currently the only facility with the capacity to manufacture fuel pellets with burnable poisons in Argentina.

2.2.1 Fuel pellets Development

The fuel pellet manufacturing process involves the following stages (Figure 1):

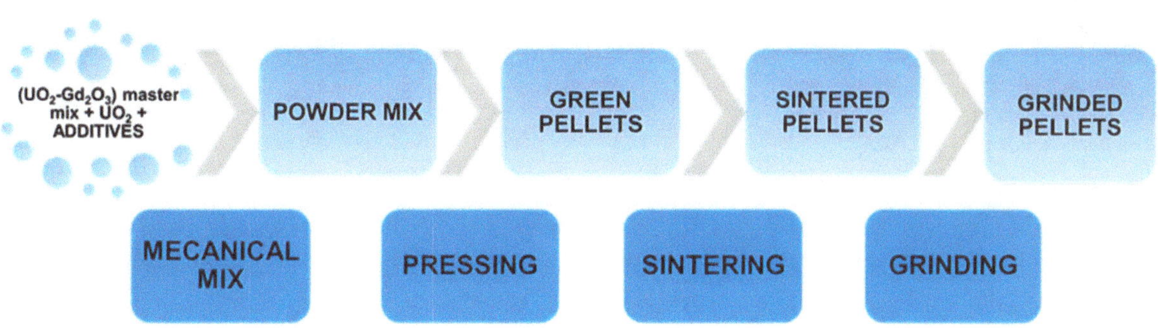

FIG. 1. UO_2-Gd_2O_3 fuel pellets manufacturing process.

Over the process, characterizations of the oxides were carried out, mixing parameters were adjusted and raw material suppliers were selected based on the homogeneity of the powders and pellets. This resulted in the production of pellets of homogeneous composition.

The compaction process of the powder mixture was done with an automatic press with a combined mechanical-hydraulic drive, defining the pressing parameters. The green pellets, obtained from the compaction, were sintered in high temperature furnaces (up to 1800°C) under a reducing atmosphere. The sintering times and temperatures were adjusted according to the results of density and thermal stability of the fuel pellets. To obtain the pellets within the planned dimensional tolerances, they were ground in a centreless cylindrical grinder. In every stage of the process, controls (visual, dimensional, density determination, etc.) were implemented and the sintered and ground pellets were subjected to QC, verifying compliance with physical, chemical, and nuclear requirements.

The objective of the development was to obtain fuel pellets that meet specifications and can be manufactured on an industrial scale, compatible with Argentine technology.

The initial development consisted of powder characterization (UO_2 ex-AUC powder and Gd_2O_3 powder). A specification was prepared for the Gd_2O_3 based on the characterization and selection of the powders from different suppliers. In Figures 2 and 3 we can see the microstructure of UO_2 ex-AUC powder and Gd_2O_3 powder.

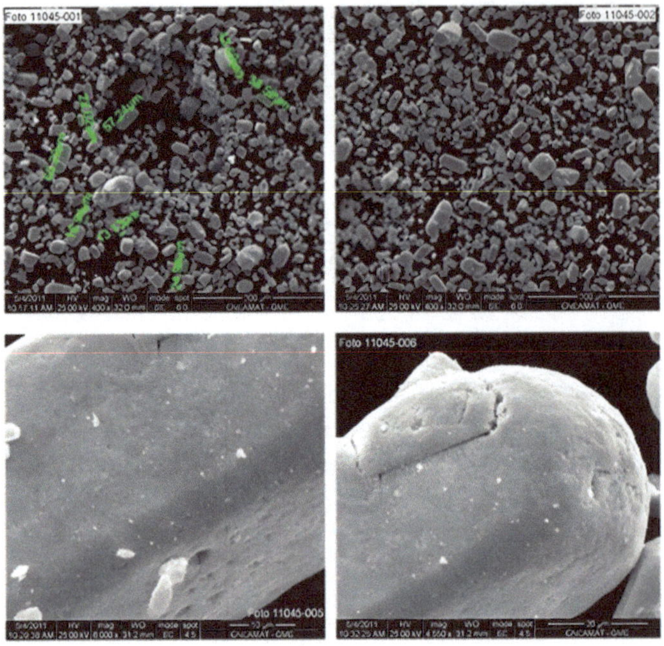

FIG. 2. UO_2 ex-AUC powder.

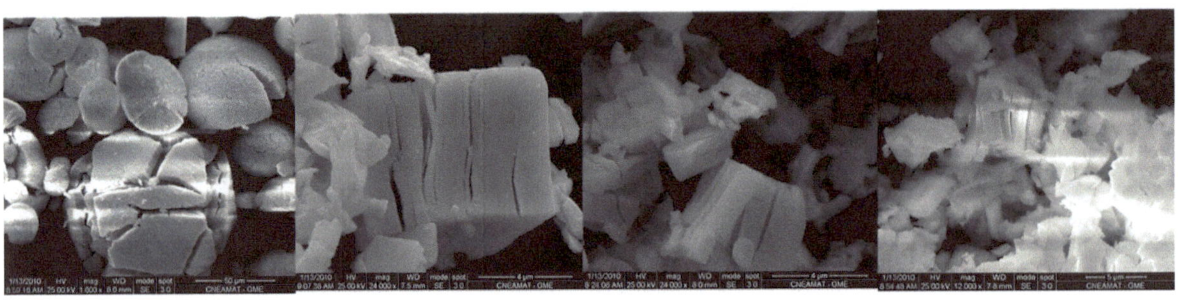

FIG. 3. Gd_2O_3 powders.

The initial tests were carried out in a development oven at 1650°C for 2 hours in an H_2 atmosphere. Working under these conditions, fuel pellets with low densities and gadolinium islands in the microstructure were obtained (Figure 4).

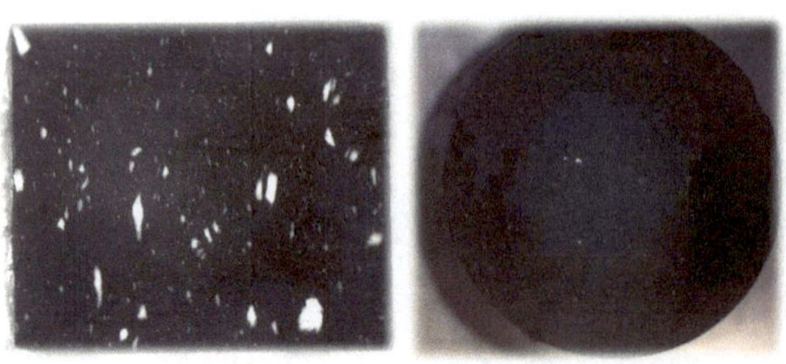

FIG. 4. Gd islands on UO_2 – Gd_2O_3 fuel pellets.

In order to resolve the microstructural problems of the fuel pellets obtained in the first tests, some process adjustments were made.

Regarding the mixing operation, we sought to optimize the process by working on the different variables involved:

- Tests were carried out with the addition of different additives with Al to improve density. Mixtures of Gd_2O_3 with $Al(OH)_3$ were made by direct mixing and co-grinding, using balls and a mortar. Mixtures were also made using aluminium stearate as a lubricant, replacing zinc stearate, opting for this alternative to incorporate Al into the fuel pellets;
- On the other hand, tests were carried out varying the mixing speed, time, mass, inclination and sequence. It was decided to carry out the mechanical mixing in two stages: master mix and dilution with UO_2.

Regarding the pressing operation, pellet shrinkage studies and a preliminary design of the tooling were carried out.

Regarding sintering, the operating conditions were modified. Another sintering furnace was used with an operating temperature of 1750°C for 4 hours in a reducing atmosphere. In this way, homogeneous tablets with thermal stability were obtained.

As a milestone to highlight within the development stage, 200 UO_2–Gd_2O_3 fuel pellets were manufactured to be irradiated in the Halden reactor (Norway). The manufactured pellets met the chemical and nuclear requirements of the specifications proposed by the CAREM25 project.

From the microstructural characterization of the fuel pellets, where the existing phases and the distribution of grains and pores were studied (Figure 5), the presence of small cracks was observed (Figure 6).

It was decided to sinter in a slightly oxidizing atmosphere to avoid the presence of small cracks. To this end, several alternatives were evaluated:

- Sintering in wet H_2 atmosphere;
- Sintering in H_2–CO_2 atmosphere. For this purpose, a test manifold was assembled and a mass flow controller for H_2 and CO_2 in different ranges was purchased;
- Increase the working temperature to 1800°C.

Finally, it was decided to carry out sintering at 1800°C in an H_2–CO_2 atmosphere. In this way, background technology for manufacturing was defined.

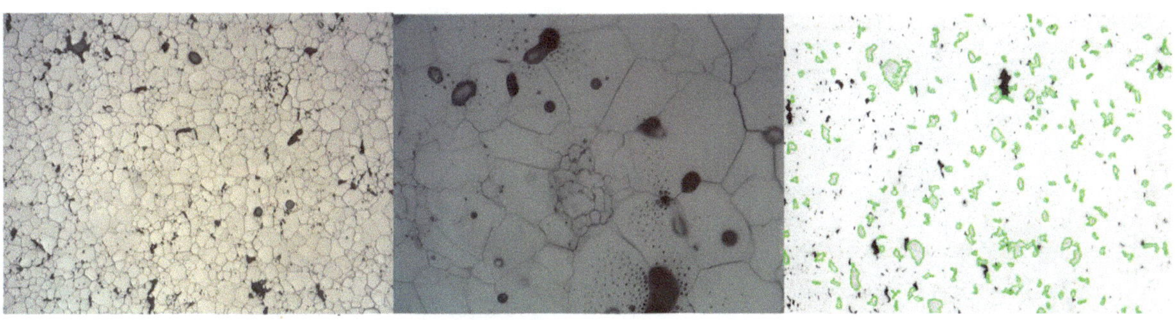

FIG. 5. Gd_2O_3 fuel pellets microstructure. The microstructure shows a distribution of grain sizes, with islands of homogeneously distributed small grains and a matrix of larger grains.

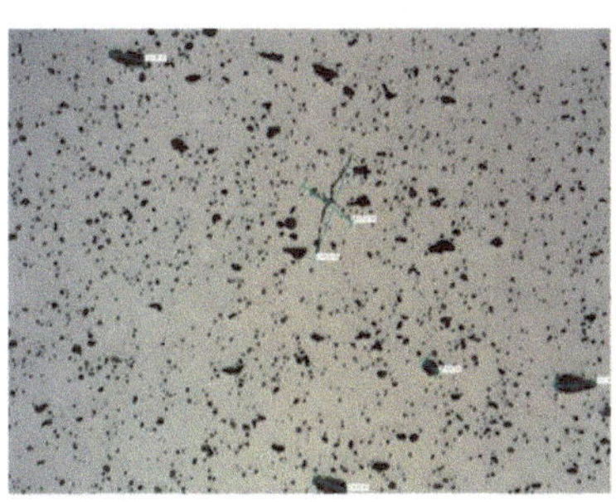

FIG. 6. Small cracks on Gd_2O_3 fuel pellets microstructure.

2.2.2 *Development and validation of characterization techniques for powders and pellets*

Regarding characterization techniques, we can classify them into three types of controls:

- Raw materials reception controls: impurity content, isotopy, humidity, stoichiometry, specific area;
- Semi-finished product controls: Mix powder homogeneity, green pellets dimensional control;
- Final controls: chemical, mechanical, dimensional and microstructural controls.

Characterization techniques were developed and validated according to the technical specifications. In addition, CNEA laboratories were qualified to carry out the tests.

In particular, it is of interest to highlight the development of resintering and colour etching ceramography. Regarding the redensification tests on UO_2–Gd_2O_3 fuel pellets, the first tests were carried out for 24 hours in an H_2 atmosphere, obtaining negative or zero redensification results. For this reason, the test method was adapted, replicating the sintering atmosphere conditions, and obtaining expected resintering values.

Regarding colour etching ceramography, it is a technique applied to homogeneity analysis in UO_2–Gd_2O_3 fuel pellets. The technique was perfected and good results were obtained (Figures 7 and 8).

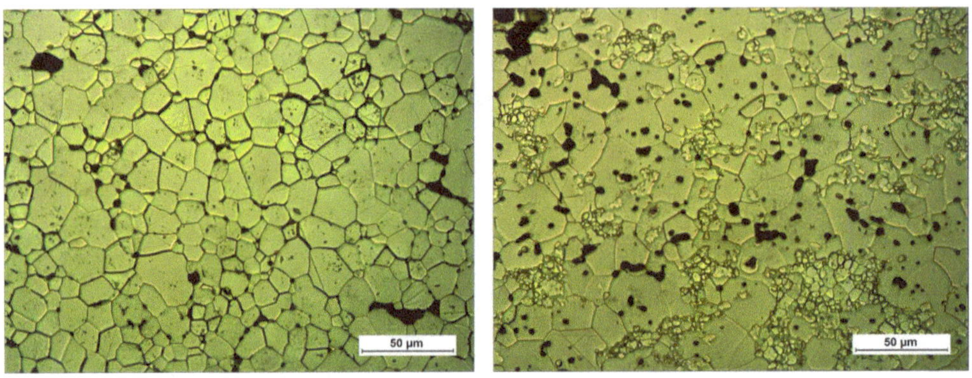

FIG. 7. UO_2 grain boundaries revealed by etching, 400x (left). UO_2 + Gd_2O_3 grain boundaries revealed by etching, 400x (right).

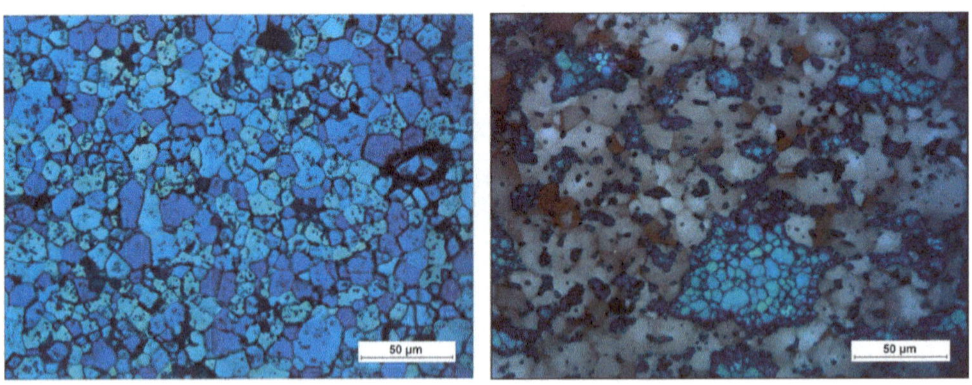

FIG. 8. UO_2 revealed by etching, 400x (left). UO_2 + Gd_2O_3 revealed by etching, 400x (right). Non-reactive UO_2 can be observed in blue.

2.2.3 *Change of scale and setting up equipment*

Once the development stage was completed, a change of scale stage was undertaken to validate the results obtained on a laboratory scale and evaluate the feasibility of manufacturing the fuel pellets on an industrial scale. This stage involved several tasks, among which the following stand out:

- Selection, acquisition, setup and adjustment of equipment.

In this sense, a three-dimensional mixer was acquired, with a capacity of up to 17 litres for the mixing operation. For the pressing operation, a load lifting system and a powder dosing system were incorporated for operation with automatic loading. Regarding the grinder, an improvement was made to the pellets discharge system and the matrix loader guide system. For the visual inspection of pellets, equipment with a rotation system was incorporated:

- Restructuring and refurbishment of facilities.

The expansion of the facility for the manufacture of fuel pellets was carried out. An area for powder treatment was added and extractor hoods connected to the ventilation system and filters were incorporated:

- Change of scale, repeatability check of outcomes with respect to the development stage;
- Definition of final parameters in the manufacturing and control processes;
- Drafting of documents for qualification (working instructions, technical reports, etc.);
- Check of the installed capacity to produce — at required scale — fuel pellets according to specifications: manufacturing of prequalification batches with the expected outcomes.

2.2.4 Qualification and manufacturing of fuel pellets for a SMR

Once the change of scale stage was satisfactorily completed, in which a repeatability check of outcomes with respect to the development stage was carried out, the process qualification stage began in view of the manufacture of UO_2–Gd_2O_3 fuel pellets for the reactor.

This stage began after a period of inactivity in the facility as a result of the COVID-19 pandemic, so additional adjustments to equipment and processes had to be made. The tasks involved in this stage were:

- Refurbishment of facilities and equipment to qualify the manufacturing process of UO_2–Gd_2O_3 fuel pellets after a period of inactivity as a result of the pandemic;
- Survey and revalidation of analytic techniques, and dust and fuel pellets characterization;
- Manufacturing of batch testing for the adjustment of operations. Characterization and analysis of outcomes. Evaluation and fitting of process parameters;
- Organization of resources for the manufacturing and characterization of fuel pellets;
- Beginning of the process qualification stage.

3. CONCLUSIONS

The development and scale-up stage of uranium dioxide– gadolinium oxide pellets (UO_2–Gd_2O_3), has been satisfactorily completed. The results obtained let us have the basic technology for the manufacture of UO_2 fuel pellets with burnable neutron poisons.

For the provision of fuel pellets for the SMR reactor, the Department carried out a survey and update on the characterization techniques involved in the process and implemented a quality system relevant to the project's requirements.

The facility is currently in the qualification stage of the manufacturing process of uranium dioxide fuel pellets with gadolinium oxide.

ID#8

DEVELOPMENT STATUS OF HIGH THERMAL CONDUCTIVE ATF PELLET AND BURNABLE ABSORBER FUEL PELLET

D-J. KIM, D-S. KIM, J. H. YANG
LWR Fuel Technology Research Division,
KAERI,
Daejeon, Republic of Korea
Email: djkim@kaeri.re.kr

Abstract

KAERI has been developing ATF pellet technologies with enhanced thermal conductivity. The fuel pellet concepts of metallic microcell and microplate UO_2 are designed as ATF pellets. Fabrication technologies for high thermal conductive fuel pellets have been developed. The thermophysical and mechanical properties of high thermal conductive fuel pellets have been measured. Out-of-pile performance tests, evaluations, and computational analyses of the fuel pellets have been conducted under various conditions and circumstances. In-reactor fuel performance data for the high thermal conductive fuel pellets were obtained and verified by irradiation tests. Additionally, high thermal conductive burnable absorber fuel pellet technologies are actively being pursued for application in SMRs, especially water cooled types.

1. INTRODUCTION

KAERI's ATF pellet technologies aim to increase operational and safety margins during normal operation and accident conditions by decreasing fuel temperature and reducing radioactivity release outside the fuel. KAERI has been developing high thermal conductive UO_2 fuel pellets, such as metallic microcell fuel and metallic microplate fuel, as part of ATF pellet technology for LWRs [1-5]. These high thermal conductive fuel technologies can also be applied to SMRs, especially water-cooled types.

High thermal conductive UO_2 fuel can significantly reduce fission gas release and rod internal pressure, which are major aspects of fuel performance and safety. It can also decrease the diffusivity and mobility of fission products and reduce fuel pellet thermal stress by lowering the fuel pellet temperature and temperature gradient [6-7].

Lower temperature fuel can decrease the radial deformation of fuel pellets under operational power transient conditions due to reduced thermal expansion. Additionally, the high ductility of the metallic phase in the fuel pellet compared to UO_2 facilitates faster creep deformation of the fuel pellets at high temperatures, thereby reducing mechanical loading on the fuel cladding. These effects are expected to be beneficial for Pellet-to-Cladding Mechanical Interaction under operational power transient conditions.

Under LOCA (Loss of Coolant Accident) conditions [6], high thermal conductive fuel can lower peak cladding temperature during fuel performance. Lower temperature fuel increases the fuel safety margin under accident conditions by decreasing stored energy in the fuel. Moreover, the high thermal conductive fuel pellet has significant beneficial effects in terms of power-to-melt.

KAERI is also conducting research and development to apply high thermal conductive fuel pellet technologies to integrated burnable absorber fuel pellets for LWRs. This technology can also be applied as a fuel technology for boron-free or low-boron operating SMRs.

2. KAERI'S ATF PELLET TECHNOLOGIES

2.1 High thermal conductive metallic microcell and microplate fuel pellet

The metallic microcell and microplate UO_2 fuel pellets are being developed as ATF pellets. The concept of a microcell UO_2 fuel pellet involves thin microcell walls enclosing UO_2 grains or granules within the UO_2 pellet. The metallic microcell UO_2 fuel pellet consists of a high thermal conductive metallic phase network and the UO_2 fuel matrix. It is expected that the metallic microcell UO_2 fuel pellet will enhance fuel performance and safety under normal operating conditions as well as during transients/accidents through higher thermal conductivity and lower temperature of the fuel pellet.

Fabrication technologies for metallic microcell fuel pellets have been developed [4, 8]. Measurements of the thermophysical and mechanical properties of metallic microcell fuel pellets have been carried out. Out-of-pile performance tests, evaluations, and computational analyses of microcell fuel pellets have been performed under various conditions and circumstances [9-13]. Additionally, in-reactor fuel performance data (in-situ online measured data of fuel centreline temperature and rod internal pressure) of metallic microcell fuel pellets were obtained and verified by irradiation tests at the Halden research reactor in Norway [14].

The metallic microplate UO_2 fuel pellet, which is another concept of high thermal conductive fuel, can effectively enhance the thermal conductivity of the UO_2 pellet. A significant quantity of metallic micro-sized thin plates is uniformly distributed within a UO_2 pellet, oriented radially. The radial alignment of these metallic microplates can effectively improve heat transfer within the fuel pellet. It is expected that various options for enhancing the thermal conductivity of fuel pellets can be provided by using metallic microcells and microplate UO_2 fuel pellets.

Figure 1 shows the microstructure image of a fabricated 5 vol.% Mo metallic microcell UO_2 pellet, in which metal phases (brighter lines) are continuously connected throughout the whole pellet.

FIG. 1. Microstructure image of fabricated Mo metallic microcell UO_2 fuel pellet.

The measured thermal conductivity of the fabricated high thermal conductive microcell fuel pellet is shown in Figure 2. Thermal conductivity of 3.5 vol.% Mo metallic microcell UO_2 fuel pellets increased by about 100% at 1000°C, compared to that of reference UO_2 pellets. It was confirmed that the simply calculated temperature using the measured thermal conductivity decreased by approximately 20%, compared to that of UO_2 fuel.

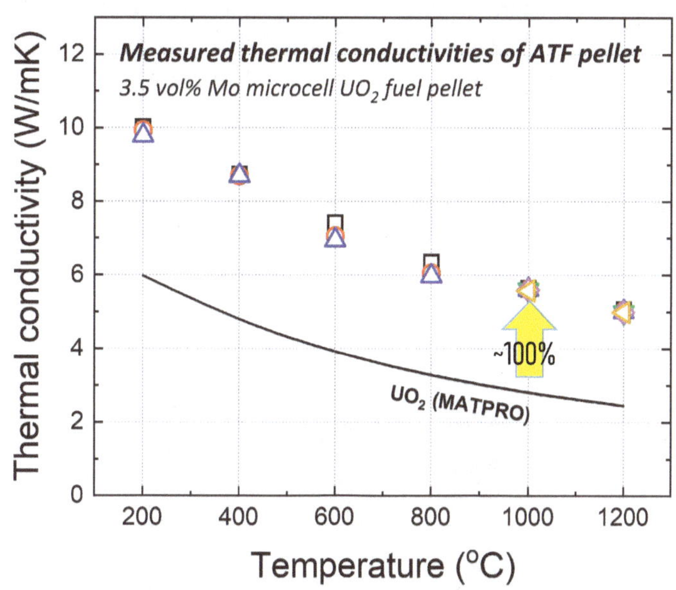

FIG. 2. Measured thermal conductivity of high thermal conductive Mo microcell fuel pellet.

Figure 3 shows the fuel centreline temperatures online measurement results in the Halden irradiation test of high thermal conductive microcell fuel pellets. The 5 vol.% Cr metallic microcell UO_2 fuel demonstrates a significantly reduced temperature compared to the reference UO_2 fuel, generally around 20% lower. These findings indicate that the metallic microcell UO_2 positively influences the fuel temperature.

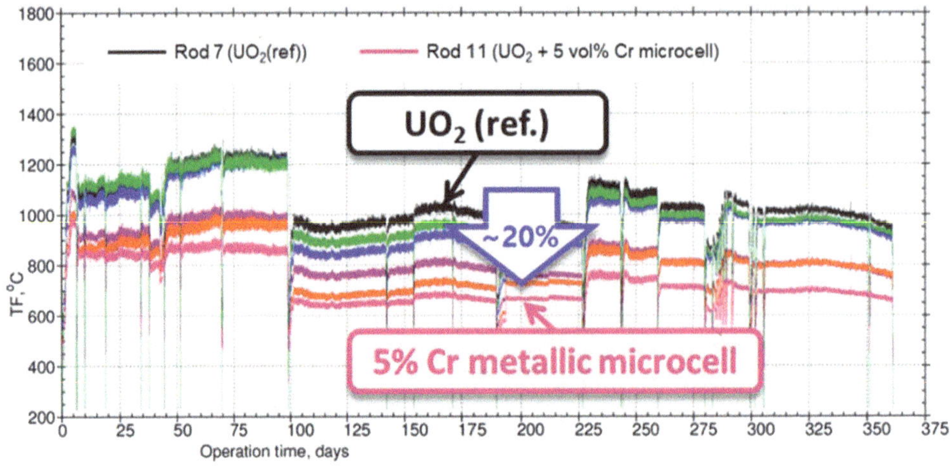

FIG. 3. Fuel centreline temperatures online measurement results in the Halden irradiation test of high thermal conductive microcell fuel pellets.

2.2 High thermal conductive burnable absorber fuel pellet

Gd_2O_3 containing UO_2 fuel pellets are typically used as burnable absorber fuel pellets (burnable poison rods) to control reactivity, especially excess reactivity, at the beginning of the cycle. However, increasing the content of Gd_2O_3 inevitably decreases the thermal conductivity of the burnable absorber UO_2 fuel pellet. This decrease in thermal conductivity results in increased fuel temperature, fuel rod power peaking, etc. Therefore, in Gd_2O_3 containing burnable absorber fuel pellets, much lower uranium enrichment needs to be applied compared to UO_2 fuel pellets.

KAERI has been developing high thermal conductive burnable absorber UO_2 fuel pellets, applying metallic microcell or microplate fuel pellet technology to integrated burnable absorber fuel pellets (see Figure 4) [17]. This approach allows for the application of higher uranium enrichment to the burnable absorber fuel pellet and is expected to be useful in applications with higher Gd_2O_3 content.

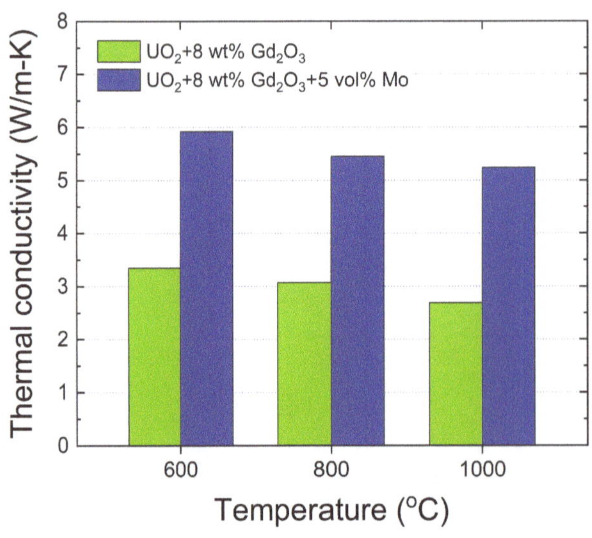

FIG. 4. Measured thermal conductivity of the high thermal conductive burnable absorber fuel pellet.

3. CONCLUSIONS

As ATF pellets, metallic microcell and metallic microplate UO_2 fuel pellets represent high thermal conductive fuels designed to improve the nuclear fuels' accident tolerance in both accident and transient conditions, while also enhancing the performance of fuel during normal operational conditions. Development has encompassed fuel concept and material design, fabrication technology, mass production process compatibility testing, and out-of-pile and in-reactor tests of higher thermal conductive fuels. Additionally, the high thermal conductive burnable absorber fuel pellet technologies are being actively performed to be applied to SMRs, especially the water-cooled type.

ACKNOWLEDGEMENT

This work was supported by the National Research Foundation of Korea (NRF) grant funded by the Korea government (Ministry of Science and ICT) (No. RS-2022-00144289).

REFERENCES

[1] KOO, Y.H., et al., KAERI's Development of LWR Accident-tolerant Fuel, J. Nucl. Technol. **186** (2014) 295.

[2] KIM, H.G., et al., Development Status of Accident-tolerant Fuel for Light Water Reactors in Korea, Nucl. Eng. Technol. **48** (2016) 1.

[3] KIM, D.J., et al., Fabrication of Micro-cell UO_2-Mo Pellet with Enhanced Thermal Conductivity, J. Nucl. Mater. **462** (2015) 289.

[4] D.J. Kim et al., Development status of microcell UO_2 pellet for accident-tolerant fuel, Nuclear Engineering and Technology, **50** (2018) 253.

[5] LEE, H. S., et al., Numerical and experimental investigation on thermal expansion of UO_2-5 vol% Mo microcell pellet for qualitative comparison to UO_2 pellet, Journal of Nuclear Materials, **518** (2019) 342.

[6] TERRANI, K. A., et al., The Effect of Fuel Thermal Conductivity on The Behavior of LWR Cores during Loss-of-coolant Accidents, Journal of Nuclear Materials, **448** (2014) 512.

[7] BROWN, N. R., et al., The Potential Impact of Enhanced Accident Tolerant Cladding Materials on Reactivity Initiated Accidents in Light Water Reactors, Annals of Nuclear Energy, **99** (2017) 353.

[8] KIM, D. S., et al., Development of Mo microplate aligned UO_2 pellets for accident tolerant fuel, Korean Nuclear Society Virtual Spring Meeting (2020).

[9] YANG, J. H., et al., Thermodynamic evaluation of equilibrium oxygen composition of UO_2-Mo nuclear fuel pellet under high temperature steam, Frontiers in Energy Research, **9** (2021) 667911.

[10] YANG, J. H., et al., Effect of molybdenum melting on the UO_2-Mo composite pellet integrity, Korean Nuclear Society Autumn Meeting, Changwon, Korea (2021).

[11] LEE, H. S., et al., Numerical investigation of thermal conductivity in UO_2-5 vol% Mo fuel pellet with Mo configuration of radial direction, Korean Nuclear Society Spring Meeting, Jeju, Korea (2020).

[12] LEE, H. S., et al., Numerical investigation of the thermal conductivity of UO_2-Mo microplate fuel pellets to realize enhanced heat transfer in the fuel radial direction, Journal of Nuclear Materials, **554** (2021) 153075.

[13] KOO, Y. H., et al., Analysis of pellet-to-cladding gap closure for a metallic micro-cell pellet under normal operating conditions, Journal of Nuclear Materials, **556** (2021) 153186.

[14] BJÖRK, K.I., et al., Irradiation testing of enhanced uranium oxide fuels, Annals of Nuclear Energy, **125** (2019) 99-106.

[15] KIM, D.J., et al., Development and various nuclear fuel applications status of accident tolerant fuel pellet technology with enhanced thermal conductivity, Korean Nuclear Society Spring Meeting, Jeju, Korea (2023).

[16] KIM, D.J., et al., Development status of high thermal conductive UO_2 pellet as ATF at KAERI, TopFuel 2019, Seattle, WA, US (2019).

[17] KIM, D.S., et al., Fabrication of integrated burnable absorber fuel pellets with enhanced thermal conductivity, Korean Nuclear Society Autumn Meeting, Changwon, Korea (2022).

ID#9

DEVELOPMENTS AND METHODOLOGY FOR THE CHARACTERIZATION OF THE MECHANICAL INTEGRITY OF PHWR FUEL PELLETS

J.P. MEDINA, L. ALVAREZ, A. BUSSOLINI
Fuel Engineering Department,
Comisión Nacional de Energía Atómica (CNEA),
Buenos Aires, Argentina
Email: medina@cnea.gov.ar

Abstract

At present there are three NPPs in operation in Argentina, Atucha 1 (CNA-1) which began operating in 1973, Embalse (CNE) which was connected to the grid in 1983 and Atucha 2 (CNA-2) which was started up in 2014.

The Argentine PHWR's have different designs; CNA-1 and CNA-2 are pressure vessels with vertical fuel channel designs while CNE is a CANDU-6. The fuel elements for commercial NPPs have undergone continuous improvement over the years, driven by operational experience, developments in fabrication techniques, and the necessity of addressing technical and economic requirements. In this regard, an important modification that can be pointed out is that CNA-1 began to use SEU in the year 2000, while the other two NPPs are still fuelled with natural uranium. The FAs and their fuel pellets have been entirely manufactured in Argentina since 1982.

The National Commission on Atomic Energy (CNEA), through the Fuel Engineering Department (IEC), is responsible for the design and engineering development of the FAs. It also provides engineering services and technical assistance to the manufacturer of the nuclear fuels. These services include the qualification of special manufacturing processes and resolution of non-conformities which may appear during the manufacturing of the three types of FAs and their intermediate products.

Regarding the manufacturing of the fuel pellets, the fabrication techniques and the controls performed in Argentina for the characterization of its chemical and physical properties are standard and similar to those employed by other manufacturers worldwide. Regarding the mechanical integrity of the pellets, there are no uniform criteria among the different manufacturers on how to characterize and to control this property during normal production. For this reason, CNEA has developed its own methodology to assess this important property.

This paper is mainly focused on explaining the methodology used by IEC to characterize the mechanical integrity of the fuel pellets, which consists of a detailed and comprehensive analysis of information obtained from different sources: Ceramographic Examination, Compression Test, Visual Inspection, Percussion Test and, in CANDU fuel, Loading and Rotation Test. It also includes a brief description of the Percussion Test and Loading and Rotation Test for CANDU fuel, which are currently in the development stage. The aim of these tests is to simulate the loading and transportation conditions of uranium pellets in the fuel rods and their impact on fuel integrity. Some application examples are also presented.

1. INTRODUCTION

The application of nuclear energy for power generation began in Argentina in 1973 with the startup of Atucha 1 NPP (CNA-1). Later, in 1983, the second NPP identified as Central Nuclear Embalse (CNE) was connected to the grid. The construction of Atucha 2 NPP (CNA-2), the 3rd NPP of Argentina, started in the 80s, halted in the 90s, relaunched in 2006, and was finished in 2014. In 2019, CNE was reconnected to the grid after the life extension tasks. Nowadays the electrical energy provided by the nuclear industry with the three NPPs in operation is about nine percent of the total electricity produced in the country.

1.1 Atucha 1 and Atucha 2

Atucha 1 and Atucha 2 are Siemens/KWU PHWR designs with pressure-vessel, moderated and cooled with heavy water (D_2O) and on power refuelling.

CNA-1 is a 357 MWe NPP. The reactor core consists of 253 FAs located in the same number of vertical coolant channels. The design of CNA-2 incorporates most of the design features of CNA-1 and lessons learned, featuring an enlarged structure and enhanced power capabilities. It generates 692 MWe, which is nearly double the capacity of its predecessor, CNA-1. This significant increase in power is made possible through the deployment of an increased number of fuel assemblies (FAs) within the reactor core, in conjunction with the use of larger-diameter fuel pellets that allow for a higher uranium content in each FA compared to those in CNA-1.

Table 1 summarizes some key characteristics of the NPPs [1-4].

TABLE 1. ARGENTINE NUCLEAR POWER PLANTS DATA

General Operating Conditions	CNA-1	CNA-2	CNE
Thermal reactor power (MWth)	1179	2160	2109
Net electric power (MWe)	335	692	600
Average specific fuel rod power (W/cm)	232	232	246
Average discharge burn-up of FAs (MWd/tU)	11400	7500	7350
Number of FAs in the core	253	451	4560
Fuel channels	253 Vertical	451 Vertical	380 Horizontal
Refuelling / fuel shuffling	on power	on power	on power

Each FA consists of circular arrays of 37 fuel rods arranged in three concentric rings and a central fuel rod. Specific structural features are also used for each type of FA. The main FA details are listed in Table 2.

TABLE 2. ARGENTINE FUEL ASSEMBLY DATA

	CNA-1	CNA-2	CNE
Number of fuel rods per FA	37	37	37
Length	6028 mm	6028 mm	495 mm
Outside diameter	107.8 mm	107.8 mm	102.8 mm
Type of spacer grids	Rigid	Elastic	--
Number of spacer grids	15 from Zry-4	13 (12 from Zry-4 and 1 from Inconel 718)	--

Each fuel rod includes a compression spring, a gas plenum, fuel and isolating pellets, all encased within a thin-walled Zircaloy-4 (Zr-4) cladding with welded end plugs at both ends. Fuel pellets stack length is around 5300 mm. CNA-1 and CNA-2 fuel rod designs are quite similar, and both use a cladding free-standing concept.

Initially, natural uranium was used as fuel material in CNA-1. However, to improve the economy of the fuel cycle, one of the main modifications was the utilization of SEU. Modifications to the design and the introduction of fuel elements with SEU into the reactor occurred from 1995 to 2000. Some of the advantages of the use of SEU are: extension of fuel discharge burnup, extension of fuel residence time, lower frequency of on-power refuelling and fuel shuffling, reduction of FAs consumption, reduction of the spent fuel volume, and therefore lower contribution of the fuel to the cost of generation. Although CNA-2 is currently fuelled with natural uranium, a project is being developed to also use SEU.

The main characteristics of the fuel rods and the fuel pellets are listed in Tables 3 and 4 [1-4].

TABLE 3. ARGENTINE FUEL RODS CHARACTERISTICS

	CNA-1	CNA-2	CNE
Cladding material		Zircaloy-4	
Cladding outside diameter	11.9 mm	12.90 mm	13.08 mm
Active length	5300 mm	5300 mm	478.6 mm
Fuel rod length	5566 mm	5566 mm	492.5 mm

TABLE 4. ARGENTINE FUEL PELLETS DESIGN DATA

	CNA-1	**CNA-2**	**CNE**
Material	Uranium dioxide		
Shape	Cylindrical pellets with dishing at both end faces		
Enrichment	SEU (0.85 w% ^{235}U)	Natural	Natural
Density (g/cm^3)	10.60	10.55	10.60
Diameter (mm)	10.62	11.57	12.16
Length (mm)	13.0	14.0	13.0

1.2 Embalse

Embalse is a CANDU-6 of 600 MWe. Between 2016 and 2018, the life extension of CNE was carried out. The main activities were the replacement of pressure tubes, steam generators and process computers. As a result, the NPP is in condition to operate for a new 30-year cycle. A summary of the primary features of the reactor is presented in Table 1. Detailed information concerning the fuel bundles is outlined in Tables 2, 3, and 4.

2. MAIN TASKS OF FUEL ENGINEERING DEPARTMENT

The National Commission on Atomic Energy (CNEA), through the Fuel Engineering Department (IEC), is in charge of the design and improvements of the FAs. To that end, it prepares drawings, specifications and other engineering documents related to raw materials, intermediate products, and FAs for each of the NPPs.

Despite the different characteristics of the three fuel designs mentioned above, the FAs and fuel pellets in particular, have been manufactured entirely in Argentina. Additionally, Argentina produces the UO_2 powder (ex-AUC) and the Zry-4 claddings.

IEC is also responsible for the qualification of special manufacturing processes involved in the manufacturing of fuel elements. In addition, IEC evaluates the Non-Conformities Requests produced during the fabrication of the fuel elements and their intermediate products, components, and raw materials. IEC also participates in fuel performance evaluations and in fuel failure analysis.

3. MANUFACTURE, QUALITY CONTROL AND CHARACTERIZATION OF FUEL PELLETS

Regardless of the differences amongst the pellets indicated in Table 4, the fabrication is performed in a single manufacturing line and basically consists of four stages: homogenization, pressing, sintering and grinding.

The control methods used in Argentina for the characterization of the physical and chemical properties of the fuel pellets and for the approval of them for loading into the fuel rods, are standard and very similar to those employed by other manufacturers worldwide. These controls cover the following characteristics:

- Uranium Content;
- Uranium Isotopic Composition;
- Chemical Composition and Impurities;
- Equivalent Boron Content;
- Content of Residual Gas;
- Hydrogen Content;
- Microstructure;
- Density;
- Cleaning and Surface Finishing;
- Shape, Size and Roughness;
- Surface Imperfections;
- Moisture Content.

Each production batch consists of approximately 150 000 fuel pellets.

4. MECHANICAL INTEGRITY OF THE FUEL PELLETS

The assessment of the fuel pellets' mechanical integrity is very important as this property may be associated with the eventual occurrence of failures of fuel elements during nuclear service. The presence of pellet fragments (chips) housed in the gap between the pellets and the sheath may affect the mechanical interaction between both components (Pellet-to-Cladding Mechanical Interaction) [5, 6]. The missing pellet fragments are associated with stress concentration failures, due to lack of adequate support when the sheath collapses on the pellet surface because of the coolant pressure [7].

Although it has been widely discussed, there are no uniform criteria for how to characterize and control the mechanical integrity of the fuel pellets during normal production.

In Argentina, IEC is working on a procedure based on a detailed analysis of the information from the following tests:

4.1 Ceramographic Examination

This control is specifically designed to detect the presence of cracks, grouped porosity, inhomogeneous grain size distributions and any other microstructural characteristics that might affect the mechanical performance of the pellets. This control is performed on a small number of selected samples from each control batch.

4.2 Compression Test

This test involves subjecting fuel pellets to compression, in axial direction, in order to evaluate their mechanical stability. For this test, a special device is used in which the fuel pellets are tested in groups of five. The test simulates the conditions that occur during the loading of the pellets into the fuel rods. This control is particularly relevant for fuel pellets to be used in long fuel rods, such as CNA-1 and CNA-2. Despite the size of CANDU fuel, a similar control is also performed on CNE fuel pellets.

The test objective is to detect the minimum forces that generate mechanical damage to the pellets. After the test, a visual inspection of the fuel pellets is performed. That allows detection of the type and number of fragments produced during the test in correspondence with the compression forces. Visual standards are used to characterize the type of chips produced during the test.

4.3 Visual inspection

A selected sample from each control batch is thoroughly inspected to seek any imperfections on the surface of the pellets, such as cracks, end-capping cracks, end surface chipping, side chipping, porous, blisters, etc. Visual standards are used to classify each pellet as: no damage, approved, and unapproved.

Ceramographic Examination, Compression Test and Visual Inspection are requirements of the pellet's specification; therefore, all batches are subjected to these controls and have to be approved prior to loading in fuel elements.

Considering last year's production of Embalse pellets, the highest proportion of defects corresponds to end-capping cracks.

4.4 Percussion Test

During both the loading of the fuel rods and their transportation, vibrations and small relative movements between the pellets are produced. These may produce chips within the fuel rods. With the purpose of simulating the loading and transportation conditions, CNEA is developing the percussion test, which aim is to subject the fuel pellets to mechanical impacts to study their chipping behaviour.

The test consists of applying a percussion force in the axial direction to the pellet's end surfaces using a vibrating bolt. At the same time, the pellets are rotated a specified number of turns. The combination of these actions simulates the axial interaction of the pellets during the loading process.

More details of the percussion equipment under development are shown in Figure 1.

FIG.1. Percussion equipment.

Generally, this control is performed on a small number of selected samples from each control batch.

After the test, each pellet is evaluated and classified according to the damage produced during the test. The categories considered are: pellets with no damage or slight detachment of the shoulder, pellets with side chipping and breakage of the pellet. Figure 2 shows the different types of damage in pellets after percussion tests.

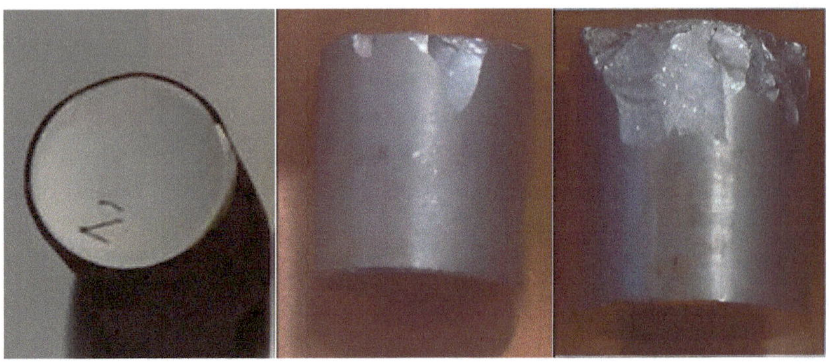

FIG.2. Different mechanical behaviour of pellets after percussion tests.

Percussion tests have been conducted on fuel pellets for the three NPPs. The results seem to reveal a relationship between the mechanical integrity and the dimensions (diameter) of the pellets.

Figure 3 shows the results of the percussion test performed on two batches of CNE pellets.

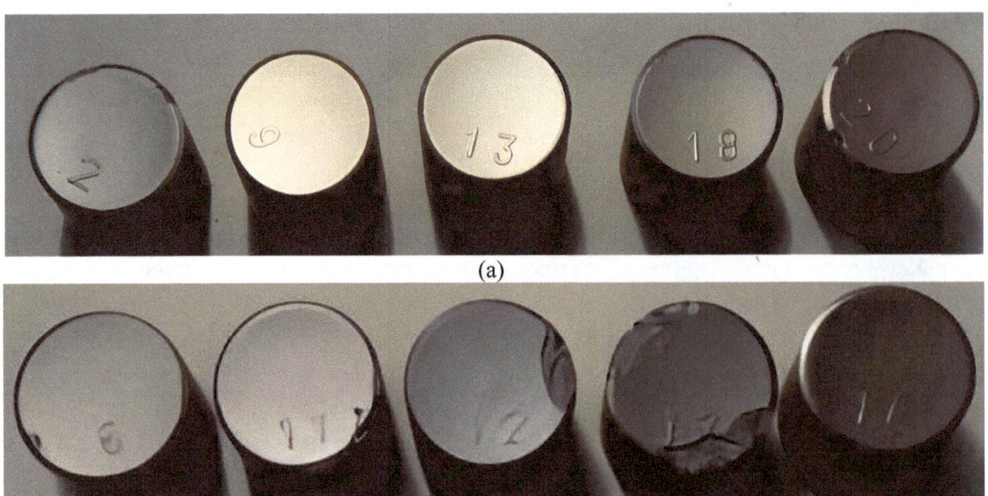

(a)

(b)

FIG.3. Results of percussion test of two CNE batches.

As an example, Figure 3 shows the difference in the mechanical behaviour of pellets corresponding to two different manufacturing batches. The pellets shown in Figure 3a show light shoulder detachment or no damage. The pellets shown in Figure 3b showed lateral breakage after the test.

Currently, this test is used only for information, and to compare the response of different batches. This procedure allows to estimate the probability of the occurrence of chips or other manifestations of structural instability in a fabrication batch.

Furthermore, considering test results, the impact of manufacturing parameters on the mechanical integrity of the pellets is being analysed.

4.5 Loading and Rotation Test

When the information obtained in the controls and characterizations previously described are not sufficient to decide about the use of the CNE fuel pellets batches, the analysis is complemented with the Loading and Rotation Test. This control is currently in the development stage. In this test, a small number of CNE fuel rods are loaded in typical manufacturing conditions and rotated for a specified number of turns.

Previously to the test, the fuel pellets to be used are inspected to detect missing pellet fragments and other surface defects that may affect the interpretation of the test result. The sample of fuel rods manufactured for this control is unloaded and inspected to detect end surface chipping and side chipping. Visual standards are used to classify each pellet. Figure 4 and 5 show patterns of each type of defect.

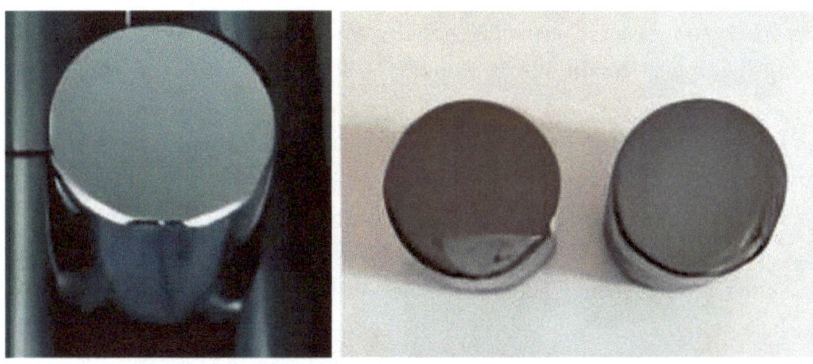

FIG. 4. End surface chipping patterns.

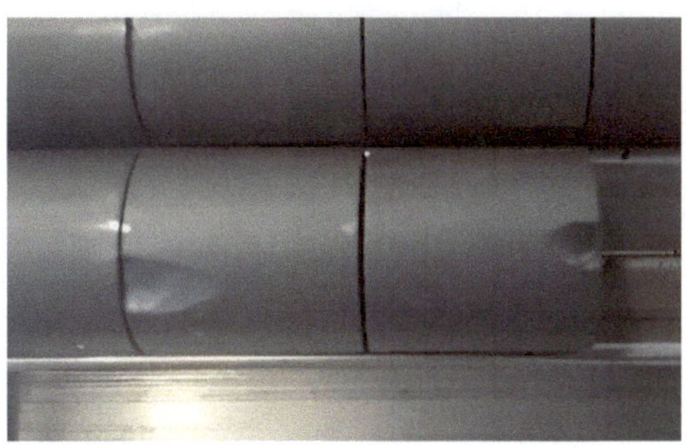

FIG. 5. Side chipping patterns.

The fragments and dust found at each tested fuel rod are also weighed and evaluated.

Ten CNE fuel pellet batches of normal production were tested. Due to the vibrations and small relative movements between the pellets during the assay, an increase in the percentage of pellets with surface defects was obtained after the Loading and Rotation Test. However, the established limit for the amount of material that can be dislodged during the test is not exceeded in any case.

5. FINAL REMARKS AND FUTURE PLANS

Fuel elements for Argentine PHWR NPPs have been manufactured in Argentina for more than 30 years.

Although the manufacturing processes of the fuel pellets are well known and the control methods used are similar to those used by other manufacturers, one of the concerns is the structural integrity of the pellets due to their sizes and other characteristics. As there is no clear global consensus about how to assess this property, CNEA has developed its own methodology for this purpose. This methodology includes a detailed analysis of the information obtained from: Ceramographic Examination, Compression Test, Visual Inspection, Percussion Test and, in CANDU fuel, Loading and Rotation Test.

In addition, CNEA is developing the Percussion Test and the Loading and Rotation test for CANDU fuel. These tests simulate the loading, manufacturing, and transportation conditions for fuel pellets to evaluate their impact on pellet integrity. Up to now, they have been used only for information and to estimate the probability of the occurrence of chips or other manifestations of structural instability. They also allow to compare results among different batches.

Although these tests may provide important information about the mechanical stability of the fuel pellets, some refinements are still required to define acceptance criteria for QC applications. Future plans include further studies with these tests to continue their development towards their inclusion in the fuel pellets specification.

On the other hand, future plans also include continued working with the manufacturer on fuel pellets manufacturing parameters or the tooling design in order to reduce end face chipping and to improve the microstructure of the pellets, particularly to reduce end capping cracks.

REFERENCES

[1] INTERNATIONAL ATOMIC ENERGY AGENCY, Review of Fuel Failures in Water Cooled Reactors (2006-2015), IAEA Nuclear Energy Series No. NF-T-2.5, IAEA, Vienna (2010).

[2] CHENG., B., Fuel Reliability Monitoring and Failure Evaluation Handbook, Revision 2, EPRI, Palo Alto, USA (2010).

[3] BÁRTA., O., Stanovení a vyhodnocenínehermetickéhopalivovéhoproutku v podmínkáchprovozu ETE (Determination and evaluation of leaking fuel rod in conditions of Temelín NPP), CHEMCOMEX Praha, a.s., Prague, Czech Republic (2003).

[4] CASARIO, J.A., ALVAREZ., L.A., Developments in slightly enriched uranium for power reactor fuel in Argentina. Impact of extended burnup on the nuclear fuel cycle, Vienna (1991).

[5] NOTARI., C., REY., F.C., Uranio levemente enriquecido en Atucha I: Un desarrollo tecnológico relevante de la CNEA y el sector nucleoeléctrico, CNEA Revista N° 3, (2001).

[6] ALVAREZ., L., et al., Status Report About Candu Type Fuel Activities in Argentina, Eighth International Conference on CANDU Fuel, Honey Harbour (2003).

[7] BUSSOLINI., A. A., et al., Atucha-2 fuel cladding design criteria and requirements for normal operating conditions. Technical Meeting on Fuel Integrity During Normal Operating and Accident Conditions in Pressurized Heavy Water Reactors, Bucharest (2012).

[8] YERMAN., J. F., Effect of fuel chips on cladding stress in Zircaloy clad oxide fuel rods. DOE research and development report (1978).

[9] SIEMENS, Siemens Nuclear Fuel Report 3 (1993).

[10] POWERS., C., et al., Hot cell examination results of non-classical PCI failures at La Salle, Top Fuel, Kyoto, (2005).

ID#10

GRAIN & PORE SIZE DISTRIBUTION OF UO2 PELLET BY CERAMOGRAPHY & IMAGE ANALYSIS METHOD'S

R. GHATREH, H. NAZEM, M. ZAHRAIE, A. RIAHI

Atomic Energy Organization of Iran (AEOI),
Iran

Abstract

The UO_2 pellet ceramography technique includes different steps such as sample preparation, measurement of the grain size distribution and average grain size, the pore's size distribution and average pore size and the pore's shape. The ceramographic analysis is conducted to verify that the microstructure characteristic of the sintered UO_2 pellet complies with the UO_2 fuel pellet specification. This study gives a full description of performing microstructural and ceramographical image analysis techniques of UO_2 fuel pellets. This technique includes sample preparation, measurement of grain size distribution, pore's size distribution and pore's shape steps by image analysis software. The pellet microstructure was successfully etched by chemical and thermal methods. The UO_2 fuel pellet microstructural has a significant effect on the physical properties such as density, densification, swelling, creep rate, thermal shock resistance, plasticity deformation and gas release. This study provides a description of the ceramographic techniques to prepare UO_2 sintered pellets in order to obtain the final statistical histogram of UO_2 pellet microstructure texture. The microstructural analysis after successful ceramographic preparation needs to be done by suitable image analysis software such as CLEMEX software.

1. INTRODUCTION

Engineering ceramics are categorized into two distinct categories: structural and functional ceramics [1]. The ceramics microstructure is greatly affected by mechanical and physical properties. The microstructure of ceramics was characterized by microscopic methods. The manufacturing parameter of high performance ceramics greatly affects microstructure such as grain or porosity. The microstructure of advanced ceramics has been studied by suitable preparation ceramography methods. The UO_2 pellet ceramographic preparation techniques include different steps described below.

2. CERAMOGRAPHY PREPARATION AND ANALYSIS OF CERAMIC

The main procedure for the preparation of ceramography analysis is very similar to metallography analysis. The main steps of preparation are: sectioning, mounting, grinding, polishing, and etching. Each of these procedures is essential, and it is vital that all steps are optimized for optimal results. The UO_2 pellets have very brittleness, porosity, and chemical resistance that very difficult to polish them as metals. Automatic sample preparation is recommended for ceramics. To reduce the polishing pressure, it is necessary to use a special diamond grinding disc instead of abrasive material [1].

2.1 Sectioning (cutting)

Generally, the UO_2 pellets are cut by a cutting machine which is cooled by water and lubricant. The pellets have been sectioned at high speed of diamond cutting (500-5000 rpm) [1]. At high speed cutting heat energy is created on the pellet cutting surface and this high temperature during the pellet cutting process can change the microstructure of the pellet. The water and lubricant caused the decreased temperature of UO_2 pellet [1].

The UO_2 pellet can be cut in two transversal and longitudinal positions. The blade operates at a rotation rate of 2000 to 5000 rpm when processing nuclear fuel pellets with an applied load of about 5 to 10 N.

The cutting rate can be excruciatingly slow for dense ceramics at 500 rpm [2].

2.2 Mechanical Preparation (Grinding and Polishing)

The UO_2 pellet has to be polished on the cutting surface. The polishing stage has to eliminate abrasion defects which appear during the cutting process. Polishing machine can be performed in automatic mode. For grinding of pellets, silicon carbide paper is used [1]. The polishing stage is performed from hard paper to soft paper in several steps. Complete elimination of any defect on the pellet surface therefore is achieved by polishing steps. For the best removal effect during the polishing, it is needed to use diamond paste from 1,3, and 6 μm. The use of 1 μm diamond polishing results in the removal of a small amount of material so 3,6 μm of diamond paste polishing can be removed highest defect. At the final stage of polishing the pellet surface has to shine and smooth [2]. The grinding is carried out in different stages as shown in the ISO 16793 standard [3].

2.3 Ceramographic Etching

Etching is certainly a very important stage of UO_2 pellet ceramography process. The grin boundaries have been sharpness corrosion by etching methods. The grain boundary's structure has a great effect on etching methods. The UO_2 pellet grain boundary etching is very difficult because this oxide ceramic texture is very resistant to corrosion by chemical or thermal methods. Each laboratory may have different conditions for the chemical and thermal etching process of UO_2 pellets. Because the empirical method of the ceramography preparation and final etching process can have a different effect on nuclear fuel samples. The texture of the ceramic microstructure such as the shape of the grain, secondary phase, and dopant agent added to nuclear fuel, porosity structure, porosity shape and many other process fabrication have been affected by ceramography etching [4]. Tow etching methods can be distinguished: Chemical etching and Thermal etching. Thermal etching is very commonly used for oxide ceramics such as nuclear pellets. Typical etching temperatures are 150°C (270°F) below the sintering temperature of ceramics in air or atmosphere [1].

2.3.1 Thermal Etching

Thermal etching method has been extensively used in ceramic microstructure analysis, which is the safest, and most highly effective technique for revealing grain boundaries within the microstructure. The polished ceramic cross section exhibits thermodynamic instability when subjected to elevated temperatures. The grain boundaries area has lower surface energy at high temperatures which causes the atom diffusion at grain boundaries to be round. The other impurities such as oil in the microstructure region have been burned in thermal etching. Thermal etching ceramics have healed some scratches on the ceramics surface. The major disadvantage of thermal etching is that large grains have been grown by spending small grains adsorbed in large grains. Sometimes grain growth has occurred in ceramography by thermal etching. The very small, grained ceramics have not to be thermally etched because grain growth has been investigated in this microstructure, preferred to grain growth. The procedure for thermal etching procedure is commonly executed at temperatures ranging from 100 to 200°C lower than the sintering temperature, with the etching duration restricted to about one hour or less [4]. The different thermal etching methods for UO_2 pellets have been reported in technical documents and standards [2-7].

2.3.2 Chemical Etching

The grain boundaries in ceramics microstructures have been higher energy than grains. The chemical etchant can have a corrosion effect at grain boundaries. Engineering ceramics especially oxides have high resistance against corrosion by chemical etchant. The best chemical etching has resulted at high temperatures and concentrations of etchant and at longer residence times. Hydrofluoric acid has a great effect on chemical etching in oxide ceramics. The mixture of hydrofluoric acid with chromic acid as the oxidant agent has been suggested by some researchers as a successful etchant of nuclear fuel pellets. The mixture of sulfuric acid with nitric acid and peroxide hydrogen as an oxidant agent is extensively suggested by some researchers for chemical agent etching of nuclear fuel pellets. The different chemical etching methods for UO_2 pellets have been reported in technical documents and standards [2-7].

2.4 Microscopic Examination

For the analyzation of ceramic microstructures SEM or light optical microscopy is commonly used. The SEM image shows the grain or grain boundaries and pore shape or pore size. Before SEM imaging the surface of the pellets has to be coated by a gold layer because the pellet's surface is an electrical insulator. The SEM secondary electron image of UO_2 pellet microstructure can be easily processed by image analysis software. The grain size distribution analysis needs etched pellets but pore size distribution can be analysed on an un-etched sample by an image analysing software. The ceramic microstructure image of UO_2 pellets by light optical microscopy has worse quality than SEM because of the light scattering limitation by the sample [1].

After the microscopic examination, we recognized the quality of pellets prepared by thermal etching is superior to that of chemical etching. For example, the quality of grain boundary images is superior when thermal etching is employed as opposed to chemical etching. In chemical etching, some grain may be overetched or underetched, which can result in grain boundaries that are not discernible in microscopic images. Therefore, it is challenging for image analysis software to distinguish these grain boundaries. So, thermal etching is a more suitable method for image analysis, although it is very important to prevent grain growth during the thermal heating process.

2.5 Quantitative Ceramography

Ceramography serves as a technique for the characterization of the grain size, grain structure, grain boundary arrangement, and pore shape of ceramics. A variety of bulk physical and thermal properties of ceramics are connected to microstructural quantities [4].

2.5.1 Grain Size

There are several ways to determine the grain diameter in a microstructure. One approach is to calculate the circle's diameter with an equal area as the mean grain area in the microstructure image. The sphere's diameter is calculated to correspond to a volume that is roughly equivalent to the average grain volume. In this method, the true average grain size exceeds the measured grain size because some grains haven't completely been in a circle by the corner. There are hundreds of grains in cross section which is cutting that only a few grains have to surround by a circle [4].

The microstructure characteristics are extensively affected by fabricating process conditions. For example, a big grain size may be caused by a high sintering temperature or a small initial particle size. The size of the grains influences different physical and mechanical properties of ceramics such as thermal properties [4].

2.5.2 Planimetric Method.

This Planimetric method, developed by Jeffries et al. in 1917, is outlined in ASTM E 112 [8]. The measurement of average grain size is achieved by a number of grains in a micrograph with a known area circle or rectangle. The number of grains per unit area changed to an average diameter has been listed in ASTM E 112 [4].

2.5.3 Lineal-Intercept Method

This Linear-Intercept method, developed by Heyn et al. in 1903, is outlined in ASTM E 112 [8]. The intercept method has to draw a cross known length line in the microstructure, then, the grain size number is accounted for by the number of grains that cross this line. Grain in the microstructure superimposed by a cross line in multiple random directions to address the irregularities of each grain. The average diameter of grain size can be calculated by ASTM E 112 standard [4]. The circle intercept method has been devised by different studies in which that method uses the circle of known circumference to compensate for oriented microstructure. This method is used in ASTM E 112 [8].

2.5.4 Image Analysis

The image of ceramics microstructure has been digitized by computer. The video camera has taken several images from ceramic microstructural. The computer image software processes the image and array of pixels characterized with x, y, and z coordinates and a grey level. The image analysis software binarized into white (255 grey level) and black (0 grey level) backgrounds. Image processing software binarized each grain which distinguished from other grains of microstructure so grain boundaries have been successfully detected by binarization of the grain microstructure. After that with statistical measurement results distribution size can be calculated. The image analysis software can measure other ceramic microstructure characteristics such as pore shape, and pore size distribution. The image analysis has some benefits to measuring all grains, quick automation analysis, reproducible and unbiased results, measure grain from other microstructure characteristics. Image analysing has some disadvantages such as some microstructural grain boundaries can't be binarized because they are discontinued or obscured, an expensive method of analysing that needs the expertise of a skilled programmer, and errors caused by bad calibration.

Semiautomatic image analysers may require detecting all grain boundaries which operator detects the unknown grain boundaries by electronic stylus or mouse [4]. The ASTM procedures have been applied to ceramography image analysis's E1245 [9] and E 1382 [10].

2.5.5 Comparison Method.

This method is very simple and fast other than methods. The comparison method considered a micrograph of a similar material with the same magnification of known grain size. The standard micrograph with specified grain size number compared by unknown image of ceramics microstructure and the similarity micrograph grain size have been reported. This method is not a true measurement because it uses a comparison method.

2.5.6 ASTM Grain Size Number.

The grain size number G defined by ASTM exhibits an inverse correlation to intercept length or average diameter that G is a logarithm related grain size number within a unit area.

2.5.7 Grain Size Distribution.

The distribution of grain sizes within a ceramic microstructure is centred around one or more mean values because the grain in the microstructure has 3 dimensional spatial orientation. Some grains may have large or small diameters in cross sections of randomly cutting orientation so that the grain size distribution greatly affects the physical properties of ceramics. The duplex microstructure is characterized by ASTM E 1181 [11].

3. RESULTS

3.1 Image analysis software to measure the grain & porosity and other microstructure parameters

We used the CLEMEX software image analysis to measure the microstructure characteristics of UO_2 pellets. We can take out the histogram of grain distribution and an average of grain size table can be extracted by the statistical table. We can get out the histogram of pore size distribution by Feret's average diameter methods and the pore size average can be extracted by the statistical table. The shape of the pores in the UO_2 pellets can be analysed by circularity, aspect ratio, spherical diameter, circular diameter, and Feret's average diameters.

3.2 The porosity size &distribution histogram in UO_2 pellet microstructure

The UO_2 pellet was cut into transection and longitudinal sections. Each section has been polished using Naples polishing cloths. The best polishing surface of pellets has been achieved using diamond paste from 1, 6, and 15 μm. Following each polishing step, the specimen has to be rotated 90° and cleaned thoroughly. Grinding and polishing lubricants are very extensively used in ceramography. The surface of UO_2 pellet has to be polished until the cross-section of pellet is free of any scratches and another defect. Lubricant and diamond paste have a great effect on shining the cross-section of the pellet and avoiding generated heating in the microstructure [4]. Pore size distribution was analysed after the polishing step and before the etching step. The light microscope has been used for this purpose. The magnification has to be used for pore size analysis and distribution is × 100, which shows the extensive area of pore distribution in the transection and longitudinal section of UO_2 pellets. Several photos have been taken for pore size analysis. Pores are detected as black areas within the microstructure of UO_2 pellets. The pore distribution across various sections can be readily observed in the subsequent Figure 1.

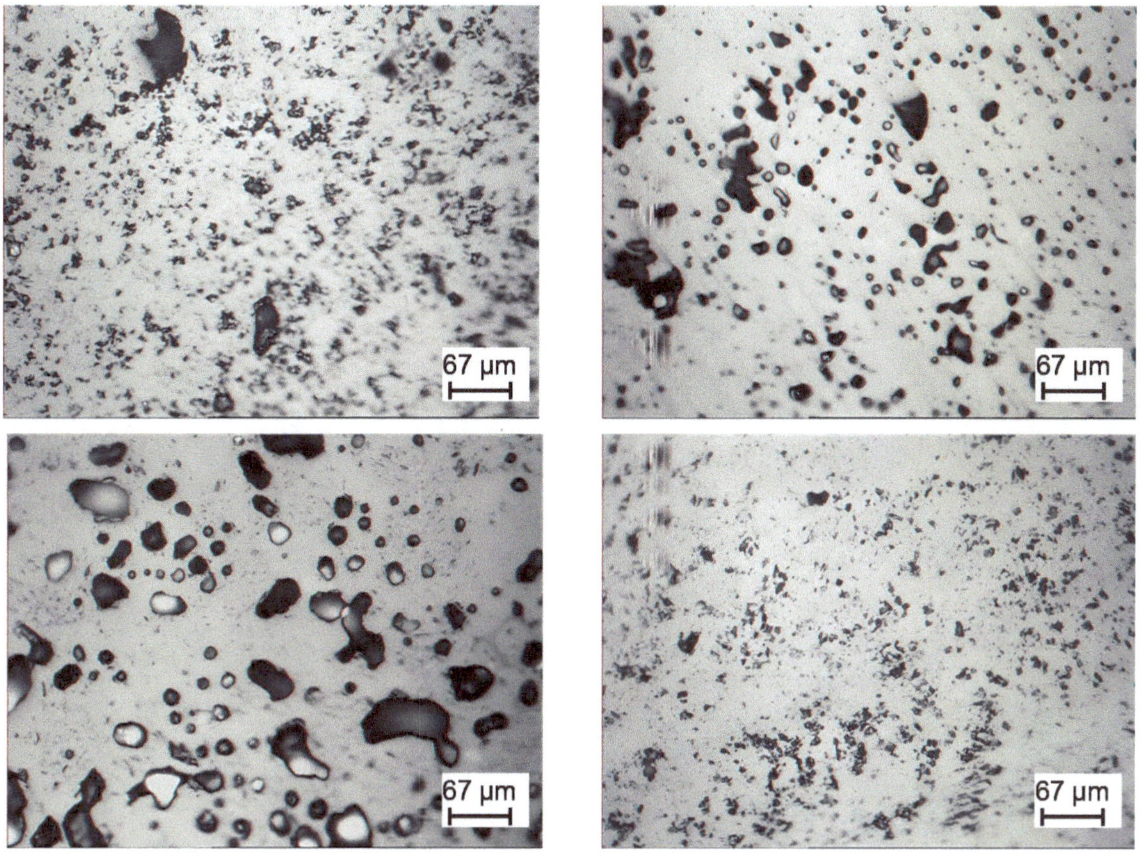

FIG. 1. The porosity image of UO_2 pellet microstructure at transection.

The images will be used by image analysis software for pore size & distribution measurement. The different functions of the software have been used for this measurement. Initially, the porosity in the image needs to be separated by threshold functions into colour in the microstructure. Usually, we separate porosity by blue colour. Subsequently, the Ferret's average and length functions have to be used to measure the porosity size. The Ferret's average size and distribution is the optimal function for measuring porosity size, which is used in nuclear fuel pellet industries. This parameter of microstructure is used extensively in nuclear fuel pellet factories to compare with specification limits. In Figure 2 the porosity size distribution in both the transection and longitudinal section of UO_2 pellets has been shown. The Ferret's size of porosity histogram needs to be compared with the specifications. For example, from the cumulative curve investigated 85% of Feret's porosity size is less than 10 μm which our specification is at least 80% of porosity size has to be less than 10 μm.

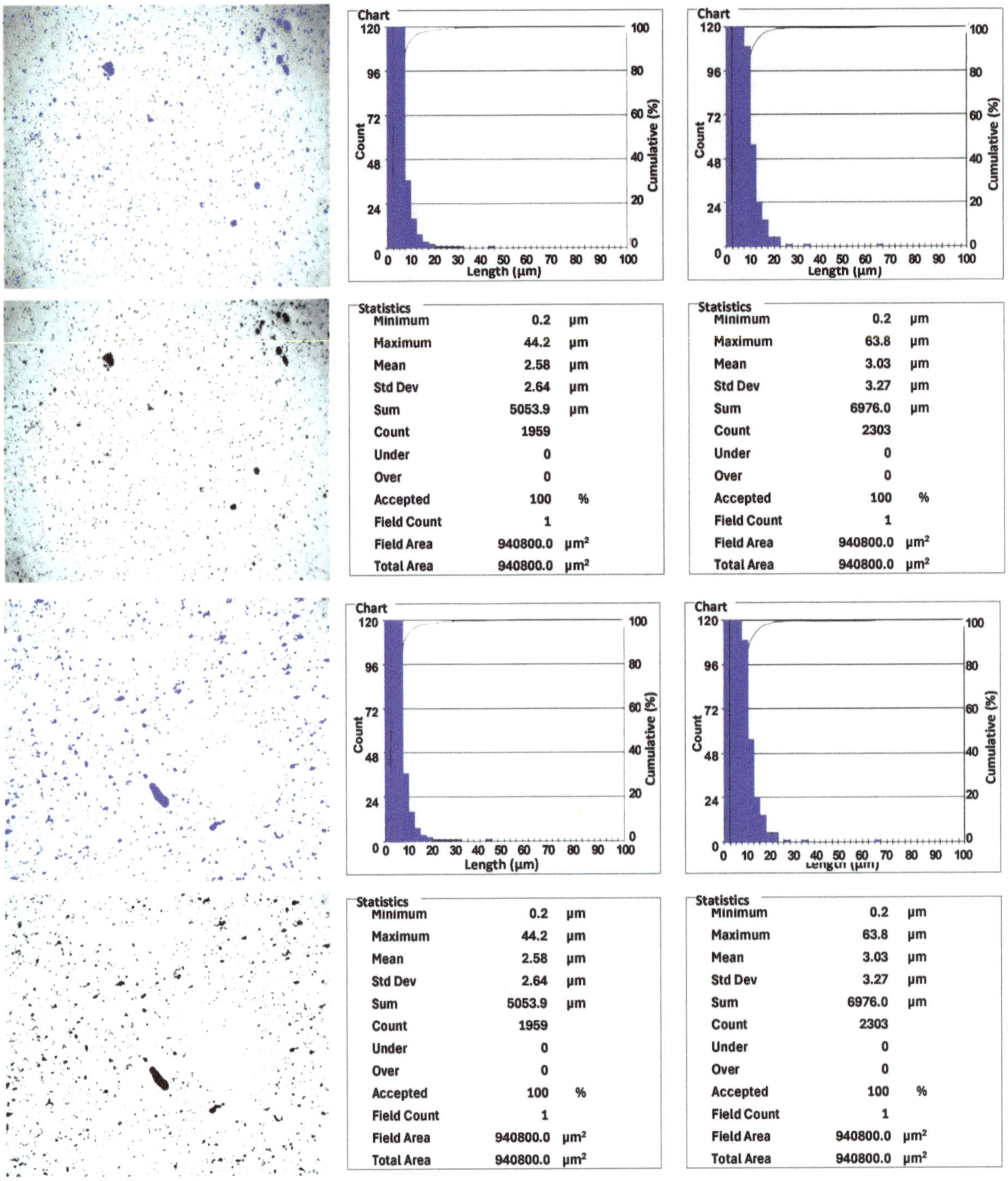

FIG. 2. The pore size distribution result of UO_2 pellet.

In the image analysis software, the different geometrical measurement histograms can be used for the measurement of porosity shape such as: Area - Perimeter Length – Width - Circular Diameter - Circularity-Roundness- Aspect ratio (L/D) – Spherical Diameter – Orientation distribution diagrams.

3.3 The grain size & distribution histogram in UO_2 pellet microstructure

After polishing and pore size measurement, the sample is etched for grain size distribution and average grain size measurement. The magnification used for pore size analysis and distribution is × 500, which shows the extensively area of grain size distribution in the transection and longitudinal section of UO_2 pellets to be captured in several photos for grain size analysis.

3.3.1 *Chemical etching*

The chemical etching agents that are typically used include acids, such as hydrofluoric acid (HF) and chromic acid (H_2CrO_4), as well as oxidants, such as hydrogen peroxide (H_2O_2) and chromium oxide (Cr_2O_3) [2]. In this procedure the sample is immersed in a solution of HF 35 vol.%, H_2CrO_4 15 vol.%, H_2O with volumetric ratio of 2:1:1 for about 2 min. During the first ½ min, the etchant with the sample still submerged is agitated by stirring or shaking. Thereafter, the sample is washed with water and dried with warm air [12]. The microstructure is delineated in this etch by differential reaction rate between grains and phases. The etching does not attack grain boundaries preferentially unless they contain impurities. A grain boundary with no impurity is observed with an electron microscope as a sloping plane between two grains of different elevations. In the light microscope, these regions appear as dark lines. The HF-H_2O_2 etch does not stain well-polished UO_2 pellets. It does, however, enlarge the pores and cavities already present, which may cause pits, particularly when the structure is deeply etched. It is crucial to note that the centre of the fuel is typically etched more rapidly than the edges, with the exception of instances where grain growth has occurred in the centre. The small grains near the edge of such a sample etch faster than the larger grains. The resulting etched surface is generally suitable for both light and electron microscopy.

Figure 3 shows the UO_2 pellet grains in the microstructure, in which grain boundaries have been successfully etched via a chemical method. It is very difficult to achieve complete etching of grain boundaries through chemical methods. However, in this study, the most effective grain boundary in the microstructure can be identified, and some grain boundaries remain unetched in the centre of the pellet microstructure. The duration of the chemical etching is a very important factor. Exceeding the recommended time may result in overetching of the grains and grain boundaries, which is not conducive to chemical etching. Therefore, the image below can be utilized for the determination of the grain size distribution by image analysis software, with some degree of error in the grain boundaries that have not been completely etched.

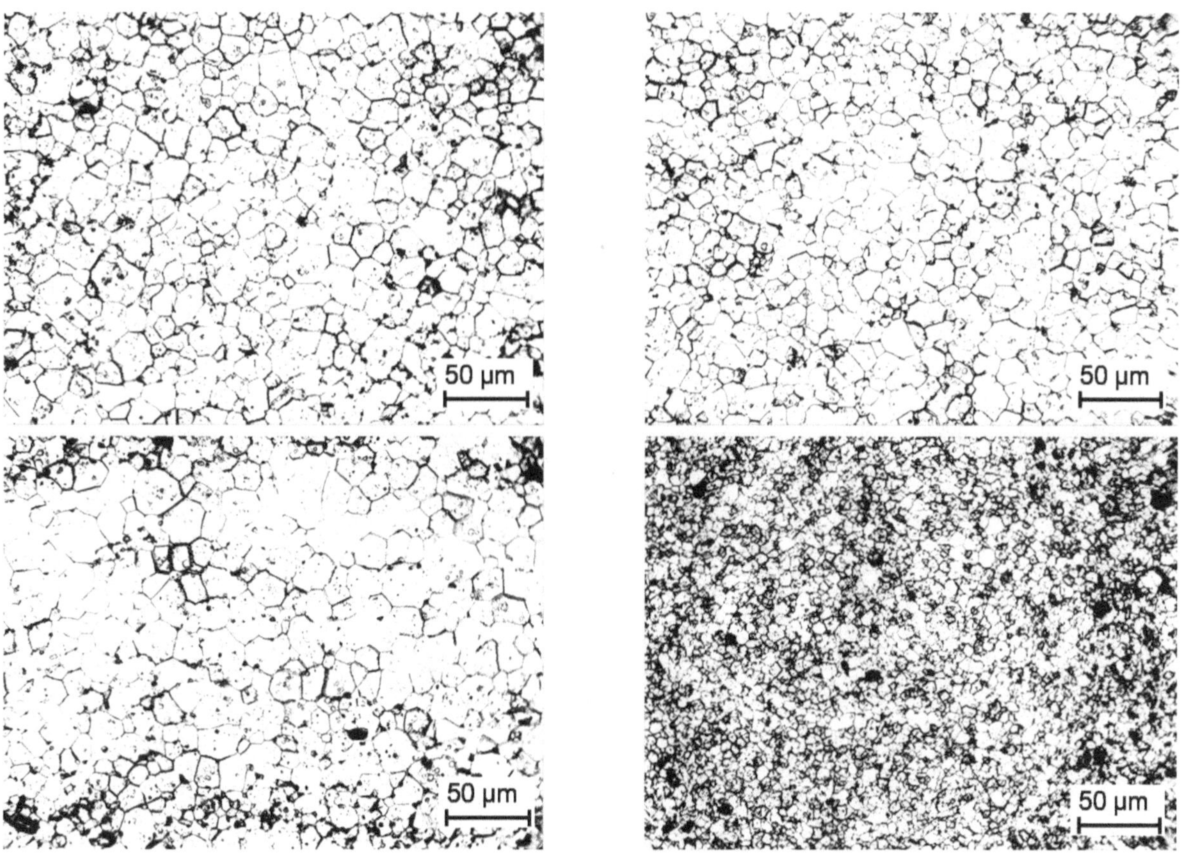

FIG. 3 The light optical microscopy image of chemically etched UO_2 pellet.

The Figure 3 images have been used in image analysis software for measuring the grain size. The different functions can be employed to asses changes in grain boundaries in the red colour. The function used for measuring the grain size distribution in the pellet microstructure is ASTM E112, after that the grain size distribution histogram has been reported by software according to ASTM E112 standard. All statistical data by the measurement of grain size reported under each histogram is shown in Figure 4. The average grain size is very important and needs to be reported. The size of grains of UO_2 pellets greatly influences the physical and thermal properties of UO_2 pellets.

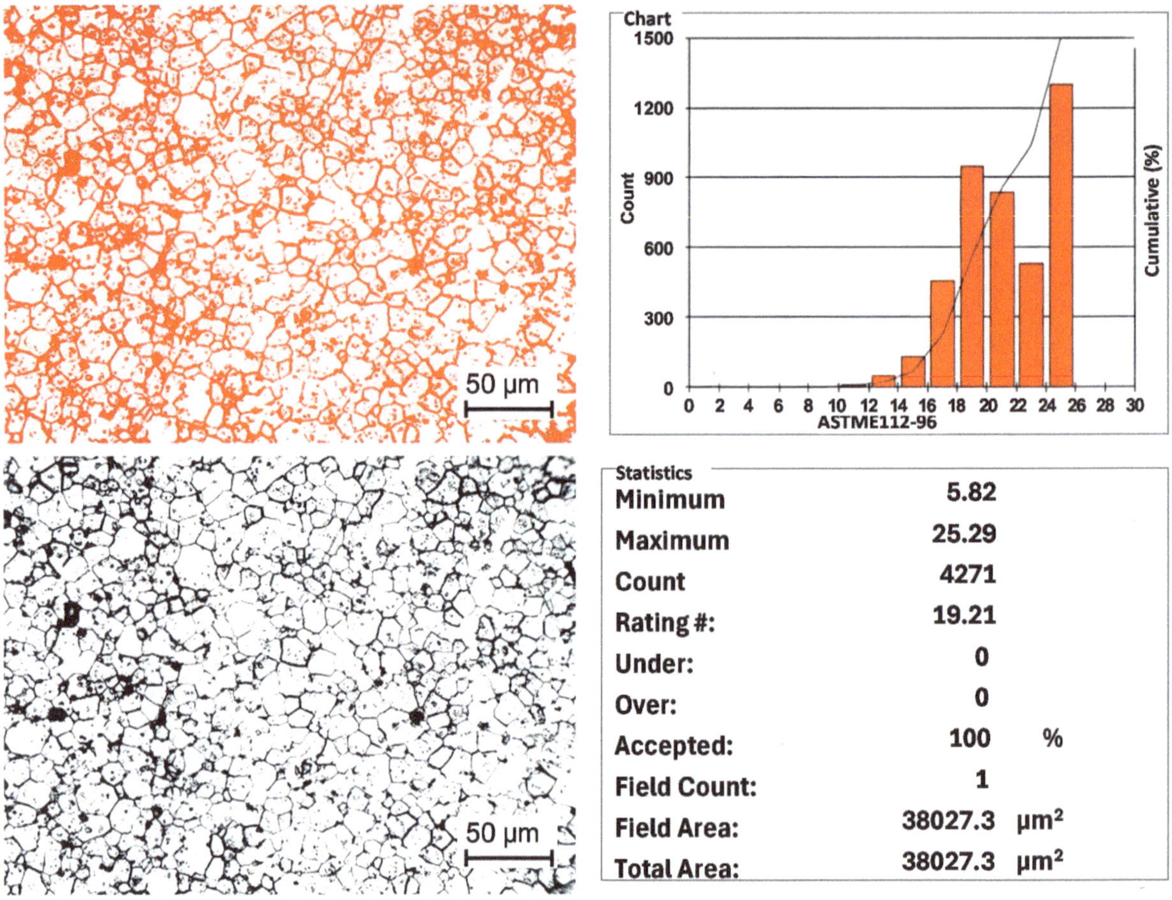

FIG.4. The grain size distribution of UO_2 pellet by chemically etched method.

3.3.2 Thermal etching

The thermal sequence varies depending on the composition of the sample and the atmosphere in the furnace. Typically, heating occurs at the highest possible ramp up to at least 1300 to 1600°C for up to approximately 5 hours [4]. For thermal etching, it is recommended to use a sectioned pellet that has been polished in a special sample holder that does not require any mounting [2]. Vector gas is preferably chosen from among CO_2, argon, other inert gases such as nitrogen and a mixture thereof. The CO_2 gas makes it possible to obtain better quality ceramographic revealing. Preferably 1300–1400°C for 30 min heat rate between 900 to 1500°C/h [13].

The different thermal etching conditions were tested on one batch of UO_2 pellets. The results are shown in Figure 5 and Table 1. According to these results, the best thermal etching condition with sharp grain boundaries in the microstructure was achieved by run #5, as shown in Figure 10. The thermal etch conditions from run #1 to run #4 resulted in an unsuitable microstructure. Some grain boundaries were overetched, while others were partially etched, with some cracks observed in the UO_2 pellet microstructures, as shown in Figures 6, 7, 8 and 9. Once the optimal microstructure condition for thermal etching is identified, the grain size analysis can be performed using image analysis software. The other position of the UO_2 pellet in the microstructure has been successfully etched. Each of the Figures was used to measure the distribution of grain sizes and the average size of grains.

TABLE 1. THERMAL ETCHING CONDITION

Maximum temperature (°C)	**Time at maximum temperature (min)**	**Heating rate (°C/min)**	**Cooling rate (°C/min)**	**Flow rate (L/min)**	**Atmosphere gas**	**Run number**
1375	45	2	5	7	CO_2	1
1300	15	2	3	7	CO_2	2
1375	15	2	Off	7	CO_2	3
1375	30	2	Off	7	CO_2	4
1400	**40**	**2**	**Off**	**7**	**CO_2**	**5**

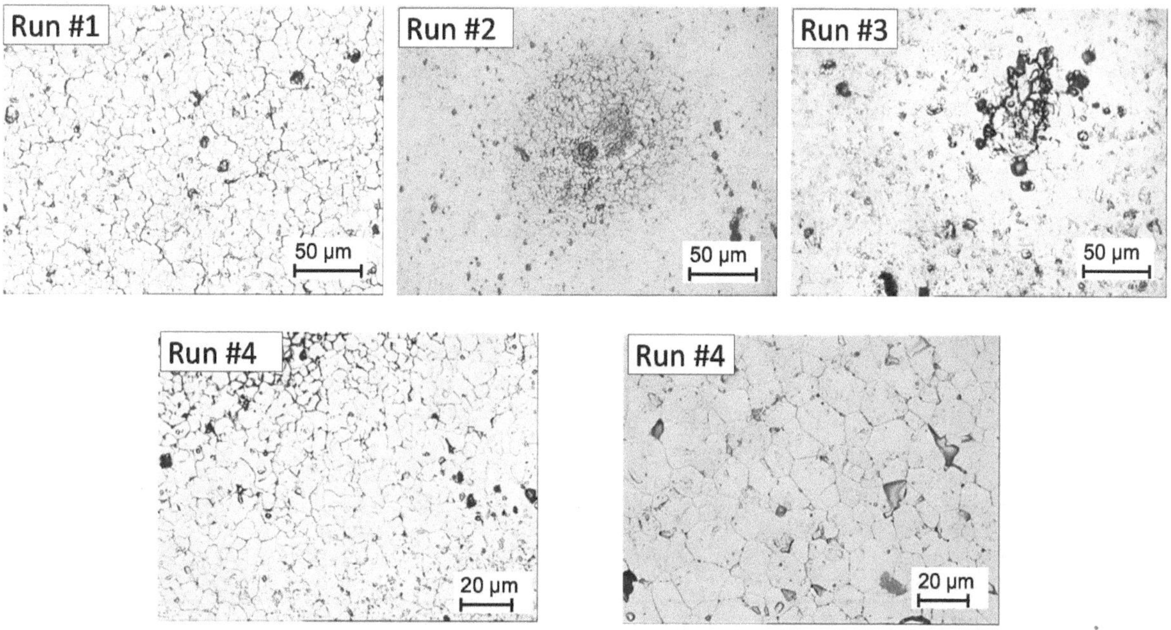

FIG. 5. The thermal etching image related to Table 1.

FIG. 6. Thermal etching run #1.

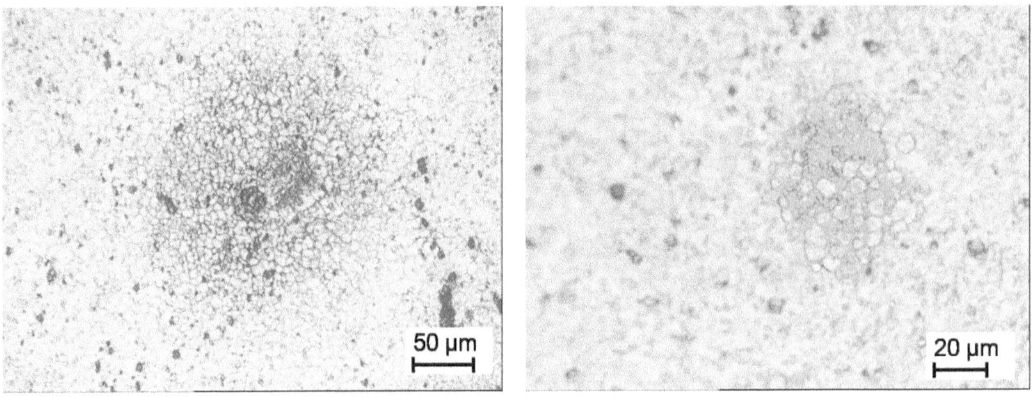

FIG. 7. Thermal etching run #2.

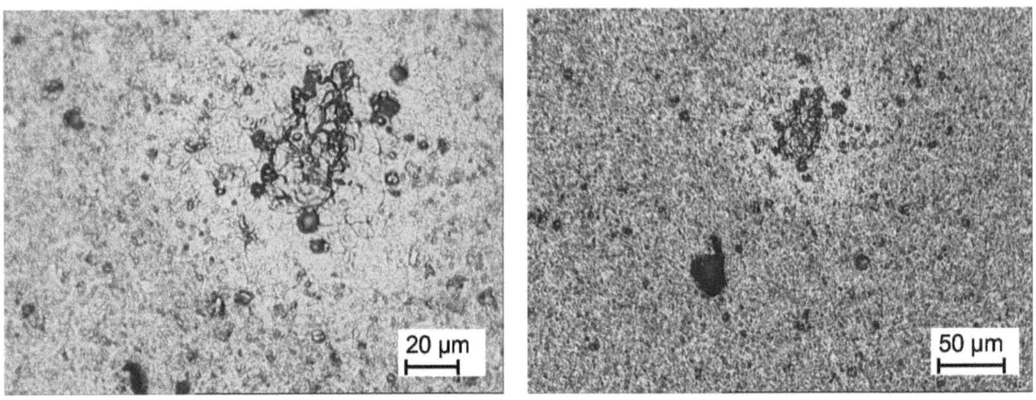

FIG. 8. Thermal etching run #3.

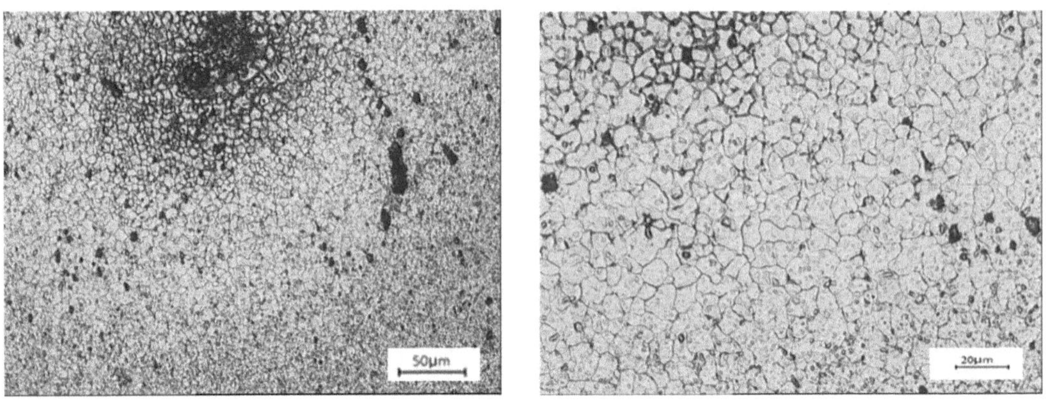

FIG. 9. Thermal etching run #4.

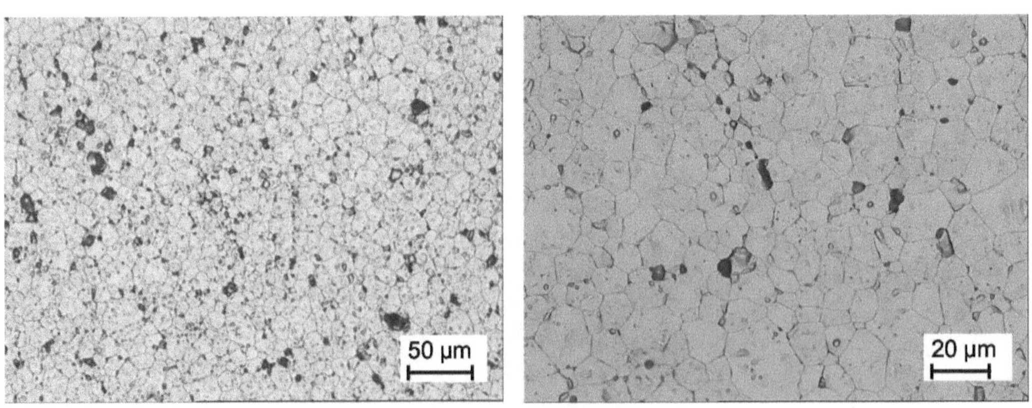

FIG. 10. Thermal etching run #5.

The result of the grain size distribution by image analysis of thermal etching preparation is shown in the Figures 11 and 12.

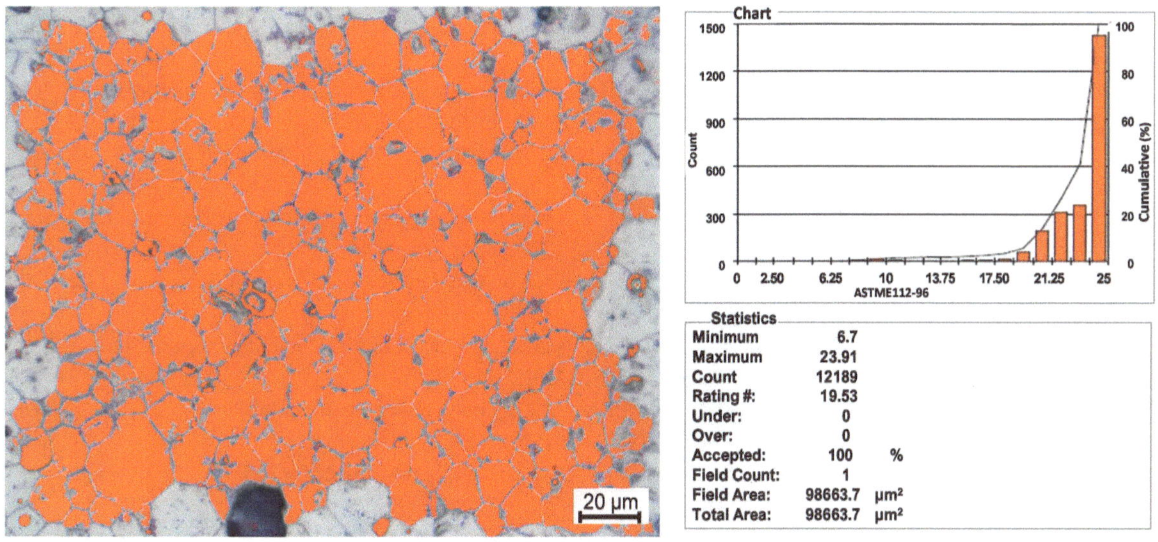

FIG. 11. The distribution of grain size & histogram 25µm.

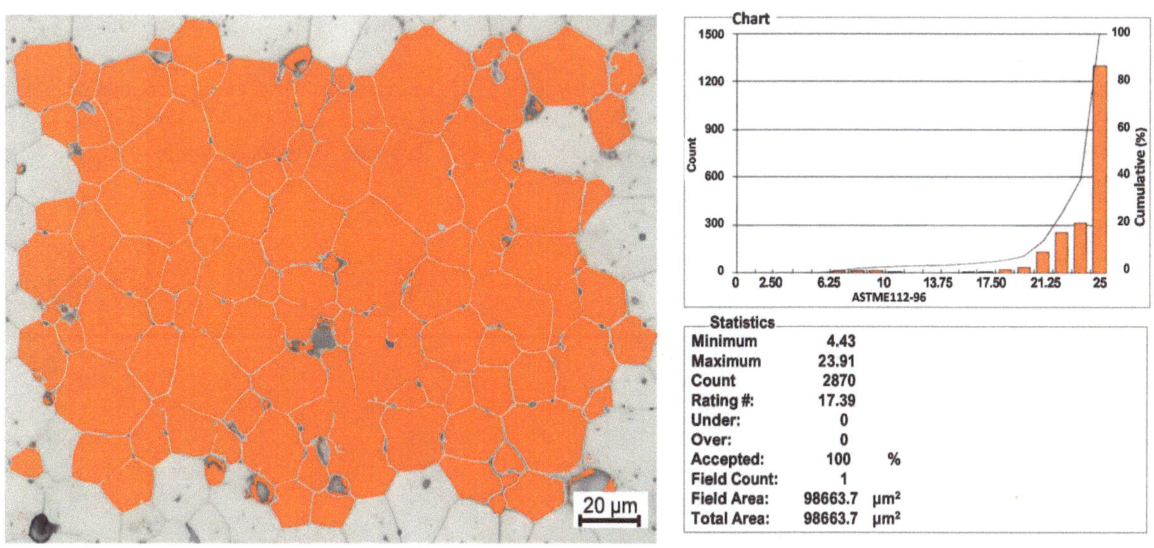

FIG. 12. The distribution of grain size & histogram.

4. CONCLUSION

The UO_2 pellet microstructure has been successfully etched by chemical and thermal methods. The best image of etched microstructure has been analysed by image analysis software for pore size and distribution, grain size and distribution measurement. The distribution histogram of pore size and grain size of UO_2 pellets is very important for fuel design performance. The pore size distribution in the range of 1–10 µm has to be more than 80% of the total pore size. The distribution of grain sizes has a great influence on the thermophysical properties of UO_2 pellets. This research has shown that the pore size distribution of UO_2 pellets between 1 and 10 µm is less than 85%. The grain size distribution is between 4 and 25 µm and the most statistical fraction of this distribution is between 17 and 25 µm, so the statistical fraction between 4 and 17 µm is very small. This method has successfully been employed to measure the microstructure properties of UO_2 pellets, as described in this research.

REFERENCES

[1] TAFFNER., U., a.U.S.f., Max-Planck-Institut fu¨ r Metallforschung, Stuttgart, Germany Michael J. Hoffmann, Preparation and Microstructural Analysis of High-Performance Ceramics, in Metallography and Microstructures. ASTM International (2004) 1057-1068.

[2] AMERICAN SOCIETY FOR TESTING AND MATERIALS, Ceramographic Preparation of UO2 and Mixed Oxide (U, Pu)O2 Pellets for Microstructural Analysis, ASTM C1868, ASTM International (2018).

[3] ISO16793., Nuclear fuel technology — Guide for ceramographic preparation of UO2 sintered pellets for microstructure examination. 2005 (E), ISO 16793.

[4] CHINN, R.E., Ceramography Preparation and Analysis of Ceramic Microstructures. 2002, Printed in the United States of America: ASM International.

[5] INTERNATIONAL ATOMIC ENERGY AGENCY, Guidebook on Quality Control of Water Reactor Fuel, Technical Reports Series No. 121, IAEA, Vienna (1983).

[6] INTERNATIONAL ATOMIC ENERGY AGENCY, Quality Assurance and Control in the Manufacture of Metal-Clad UO2 Reactor Fuels. Technical Reports Series No. 173, IAEA, Vienna (1976).

[7] AMERICAN SOCIETY FOR TESTING AND MATERIALS, Ceramographic preparation of PuO2-UO2 mixed oxide fuel pellets and uranium pellets. ASTM C-994, ASTM International (1983).

[8] AMERICAN SOCIETY FOR TESTING AND MATERIALS, Standard Test Methods for Determining Average Grain Size, ASTM E112, ASTM International (2013).

[9] AMERICAN SOCIETY FOR TESTING AND MATERIALS, Standard Practice for Determining the Inclusion or Second-Phase Constituent Content of Metals by Automatic Image Analysis, ASTM E1245, ASTM International (2016).

[10] AMERICAN SOCIETY FOR TESTING AND MATERIALS, Standard Test Methods for Determining Average Grain Size Using Semiautomatic and Automatic Image Analysis, ASTM E1382, ASTM International (2015).

[11] AMERICAN SOCIETY FOR TESTING AND MATERIALS, Standard Test Methods for Characterizing Duplex Grain Sizes, ASTM E1181, ASTM International (2015).

[12] BELLE., J., Uranium Dioxide: properties and nuclear applications. 1961: United state atomic energy commission.

[13] CHAROLLAIS., F., et al., Thermal etching process of a ceramics under oxidizing conditions. 2001, patent number US6171511B1 Commissariat a l'Energie Atomique, Paris; Compagnie Generate des Matieres Nucleaires, Velizy-Villacoublay, both of (FR).

TECHNICAL SESSION III: CLADDING, FUEL ROD AND FUEL ASSEMBLY COMPONENTS

ID#11

ALTERNATIVES TO REPLACE THE USE OF BERYLLIUM FOR APPENDAGES ATTACHMENT ON CANDU FUEL ELEMENTS

A. BUSSOLINI, M. OLIVERA MUÑOZ; A. MINETTI, L. ALVAREZ
Fuel Engineering Department,
Comisión Nacional de Energía Atómica (CNEA),
Buenos Aires, Argentina

Abstract

Beryllium (Be) is the filler material which is used in the brazing of the appendages for CANDU fuel bundles for Embalse PHWR, Argentina. Using Be has some drawbacks in terms of environmental conditions and health risks to workers, so it is intended to eliminate its utilization in the fuel element manufacturing process. Based on the local experience on resistance welding of bearing pads and tabs to the claddings for Atucha 1 and Atucha 2 fuel rods, this welding process is the main alternative to consider for the replacement and this work focuses on showing this experience and the feasibility of such replacement. Other options are also under consideration.

1. INTRODUCTION

There are three NPPs in operation in Argentina: Atucha 1 (CNA-1) which has been operating since 1973, Embalse (CNE) which was connected to the grid in 1983, and Atucha 2 (CNA-2) was started up in 2014. The net electrical power outputs of these plants are 350 MW, 600 MW and 700 MW respectively.

Although the three plants are PHWR their designs and characteristics are different: CNA-1 and CNA-2 are pressure vessel designs with vertical fuel channels while CNE is a CANDU-6. Furthermore, CNA-1 has used SEU since the year 2000, whereas the other two NPPs are fuelled with natural uranium. The FA designs for the three NPPS are also different, but the three designs feature a total of 37 fuel rods (a central fuel rod and three concentric rings). Zircaloy-4 is the cladding material.

In Argentina, the FAs for the commercial NPPs are wholly fabricated, with their designs having undergone continuous improvement over the years, driven by operational experience, developments in fabrication techniques, and the necessity of addressing technical and economic requirements.

2. CNA-1 AND CNA-2 FAs

CNA-1 and CNA-2 fuel rods are almost 6 meters long, free standing design with bearing pads and tabs welded to the cladding by resistance welding (Figure 1).

Each CNA-1 FA has 1665 bearing pads welded by resistance welding.

FIG. 1. CNA-1 fuel rods with bearing pads in contact with the spacer grid.

In the CNA-2 FA the components welded to the fuel rods, as seen in Figure 2, are named tabs and the number of tabs per FA is 224.

FIG. 2. CNA-2 FAs. Fuel rods with tabs welded to the cladding by resistance welding.

2.1 Appendages resistance welding experience

The local manufacturer (CONUAR-FAE) has been producing fuel rods with bearing pads welded to the cladding by resistance welding for CNA-1 since 1982.

The similarity between CNA-1 and CNA-2 fuel rods made it possible to apply the same welding process for tab attachment in the same production line with some differences in tools and welding parameters.

2.2 Manufacturing Process

The tabs for CNA-2 and bearing pads for CNA-1 are obtained by Zr-4 wire cutting. The manufacturing process consists of the following steps:

- Cleaning of tabs or bearing pads (ultrasonic bath) and cladding;
- Resistance welding: automatic positioning of the tabs or bearing pads along the 6-meter length of the fuel rod and welding by controlling electric current and pressure applied using an internal mandrel to reduce cladding deformations;
- Machining of tabs or bearing pads: once welded, they are machined to obtain the final geometry of tabs or bearing pads (Figure 3).

FIG. 3. CNA-1 bearing pads before and after final machining.

To verify the compliance with resistance welding requirements, the following QCs are carried out:

- Dimensional controls;
- Corrosion test;
- Metallographic inspection;
- Mechanical strength;
- Visual inspection.

3. CNE FUEL ASSEMBLY

CNE FA is a typical 37 element CANDU fuel bundle (Figure 4). Fuel rods are almost half a meter long and have collapsible cladding with appendages attached to the cladding by brazing. There are two types of appendages:

- The spacers (Figure 5a) allow to maintain separation between the fuel rods;
- The bearing pads (Figure 5b) are only in the fuel rods located in the external diameter of the fuel bundle to ensure that the cladding does not come directly into contact with the pressure tube or other internals.

FIG. 4. Typical CANDU-6 fuel bundle.

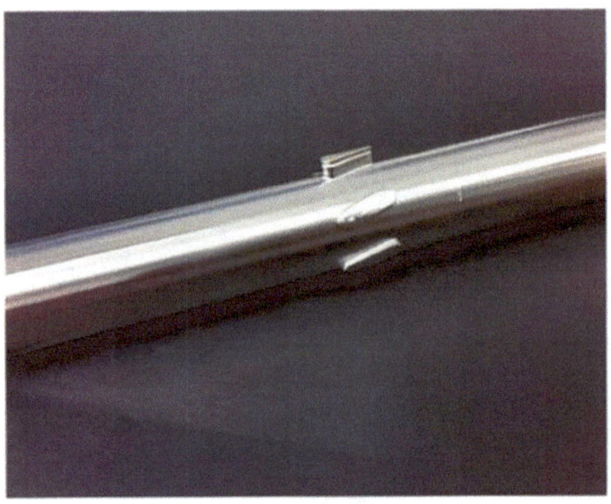

(a)

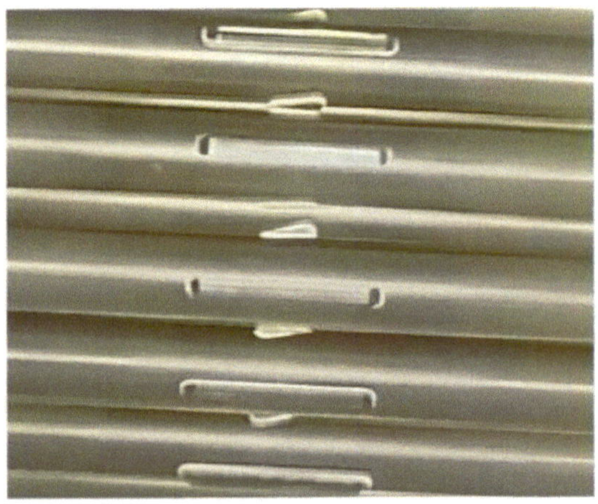

(b)

FIG. 5. CANDU fuel rod spacers (a) and bearing pads (b).

Both types of appendages are from Zircaloy-4. Beryllium (Be) is the filler material used in the brazing of the appendages to the cladding as it is the state of the art in CANDU fuel bundle manufacturing.

3.1 Brazing process using Be

The brazing process with Be starts with the metal deposition of Be in the Zircalloy-4 sheets, the formation of the eutectic composition followed by mechanical pressing cutting to obtain the appendages with a surface ready for brazing. After that, the appendages are fixed in their position by two spots of resistance welding. Finally, the brazing process is carried out by induction. In the CNE fuel bundle, there are five types of fuel rods, in some of them to get the required thickness of the appendages in the fuel rods, they are joined overlapped using the same procedure (Figure 6).

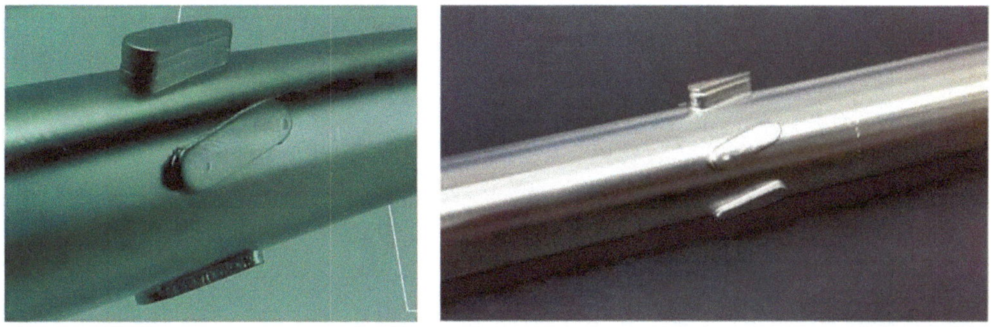

FIG. 6. CANDU fuel rod spacers.

To verify the compliance with the specified requirements, the following QCs are carried out:

- Dimensional controls;
- Visual inspection;
- Metallographic control (Figure 7);
- Corrosion test (only at the qualification stage).

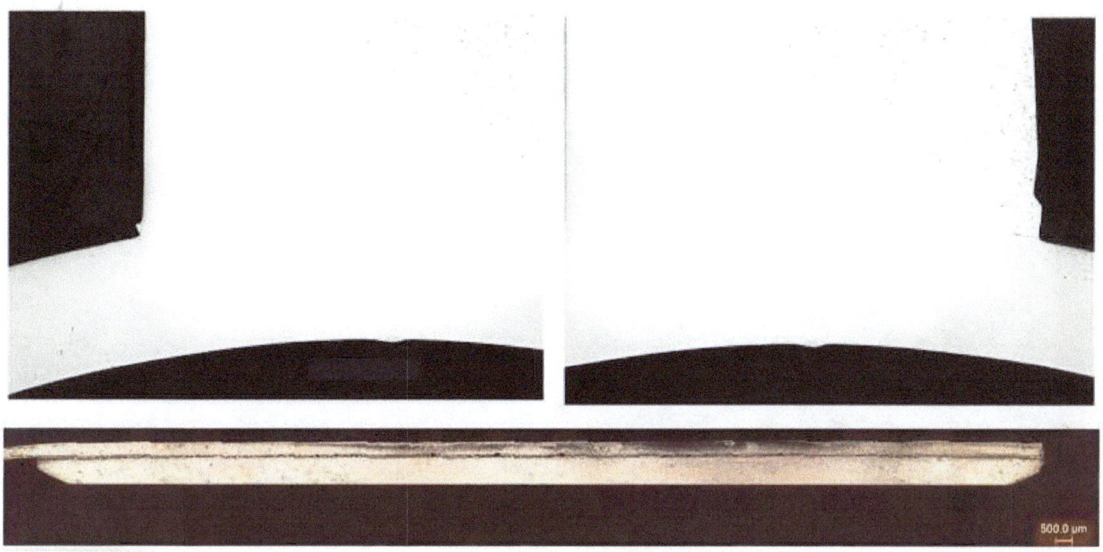

FIG. 7. Metallographic inspection of appendages welded by brazing to the cladding at CNE fuel rods.

Through the years the local manufacturer has achieved a very reliable and stable brazing process with no issues to fulfil current specifications. Despite that, the use of Be has some drawbacks in terms of environmental issues and health risks to workers. The management of this material and process waste is required to comply with strict requirements and scrap is not suitable for recycling.

4. ALTERNATIVES TO BE BRAZING

To eliminate the risks and limitations it is intended to eliminate the utilization of Be in the CNE fuel element manufacturing process. The proposed alternatives are replacing the brazing filler material or employing other welding methods.

There are several brazing materials under consideration such as aluminium, titanium or nickel–chromium based alloys.

Alternative welding methods under consideration are resistance, laser or electron beam welding.

Between the different options presented, the implementation of the resistance welding process appears to be the most suitable to be applied in the short or medium term given the available equipment and the experience and expertise in manufacturing CNA-1 and CNA-2 fuel rods. Some design differences with CNE fuel rods do not allow the direct implementation of this method and lead to the need for some changes in the welding process. The main design differences in comparison with CNA-1 and CNA-2 are:

- Cladding is thinner;
- Geometry and positioning of the appendages are more complex, as can be seen in Figures 1, 2, 3, 5, and 6;
- Overall quantity of the appendages is larger.

4.1 Resistance welding preliminary tests

The Argentine local fuel manufacturer (CONUAR-FAE) has made some preliminary tests by welding bearing pads with CNA-1 and CNA-2 resistance welding machines and new tooling to adapt the process to CNE fuel rods. New jaws, mandrel for internal holding, and electrode holders with different parameters (roughness of the bearing pads, current, and pressure) adapted for CNE fuel rods were used.

Quality test results are promising in terms of metallographic microstructure in the welded zone (Figure 8) and manual torque test (Figure 9). Despite these satisfactory results, there are still some issues to be improved in terms of uniformity along the length of the welding (Figure 10).

FIG. 8. Metallographic inspection of resistance welding zone of the bearing pad with the cladding.

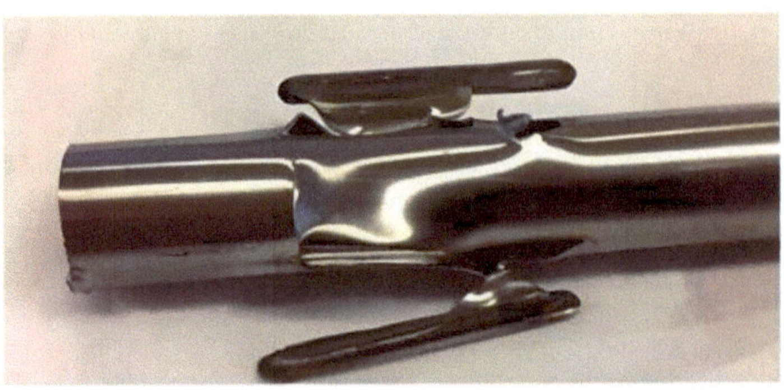

FIG. 9. Manual torque test showing no fracture in welding joints.

FIG. 10. Macro metallographic inspection of resistance welding zone of bearing pad to cladding.

4.2 Future work

An extensive road map that leads to the industrial implementation has been proposed and includes the following steps:

- Characterization plan;
- Perform verification tests and results interpretation;
- Adapt the requirements of the manufacturing and control documentation to the new processes;
- Qualification of the manufacturing processes to demonstrate that specification requirements are fulfilled and are stable;
- Define a pilot plan for a gradual implementation.

5. CONCLUSIONS

Based on local experience with resistance welding for attaching pads and tabs to the claddings, it seems to be a promising alternative to replace the Be in the brazing welding in CANDU-6 fuel rods. Although some preliminary tests show good results, there are some issues to improve to be used as the default process. Besides, as can be seen, eliminating Beryllium is not just an advantage regarding environmental conditions, it also implies an advantage during the manufacturing process and an improvement in the reliability and sustainability of the fuel manufacturing activity.

ID#12
DEVELOPMENTS IN FUEL FABRICATION AT CNL

F. DIMAYUGA, L. XIAO
Canadian Nuclear Laboratories,
Chalk River, ON, Canada

Abstract

Canadian Nuclear Laboratories has a long history of research and development in nuclear fuels in the areas of fabrication development, characterization, out-reactor testing, irradiation testing, and performance assessments through post-irradiation examination. CNL has been utilizing its extensive knowledge and dedicated facilities to advance the development of existing fuel technologies being used in existing reactor designs, as well as support the development of novel techniques and materials for advanced fuel applications. Much of this activity is performed through projects within the Federal Nuclear Science and Technology Work Plan, a programme established and managed by Atomic Energy of Canada Limited. This programme is intended to facilitate CNL's contributions to the Government of Canada's objectives in health, science, innovation and climate change. Additionally, CNL is involved in fuel development initiatives within the framework of the R&D programme financed by the CANDU Owners Group.

This paper presents a sampling of the development work on fuel fabrication technologies being conducted at CNL using two specific examples: one is a study of new filler materials for brazing of zirconium alloys, and another is the work on advanced fuel cladding alloys. Results of the out-reactor and in-reactor testing of stainless steel as a braze filler material to replace beryllium will be discussed, and results of out-reactor testing for mechanical properties of Zr–Nb will be presented. Important developments achieved in these projects are highlighted, and future direction of R&D effort is provided.

Key words: nuclear fuel fabrication, braze filler materials, advanced cladding alloys, joining techniques.

1. INTRODUCTION

For decades, the CNL have been conducting research and development in nuclear fuels in the areas of fabrication development, characterization, out-reactor testing, irradiation testing, and post-irradiation examination of various types of fuels, ranging from oxide fuels intended for power reactors to metallic fuels for use in research reactors. CNL has been utilizing its extensive knowledge and dedicated facilities to advance the development of existing fuel technologies being used in existing reactor designs, as well as support the development of novel techniques and materials for advanced fuel applications. Much of this activity is performed through projects within the Federal Nuclear Science and Technology Work Plan, a programme established and managed by Atomic Energy of Canada Limited. This programme is intended to facilitate CNL's contributions to the Government of Canada's objectives in health, science, innovation and climate change [1]. Additionally, CNL is involved in fuel development initiatives within the framework of the R&D programme financed by the CANDU Owners Group [2].

This paper presents a sampling of the development work on fuel fabrication technologies being conducted at CNL with a focus on applications to power reactor fuels, particularly CANDU fuels. One study that will be discussed is the investigation of new filler materials for brazing to replace beryllium as applied to zirconium alloys. Some of the work on Zr–Nb, an advanced fuel cladding alloy, is also discussed, particularly in terms of results of out-reactor testing for mechanical properties that can inform their potential suitability to high-burnup or ATF applications. Important developments achieved in these projects are highlighted, and future direction of R&D effort is provided.

2. NEW FILLER MATERIAL FOR BRAZING OF CANDU FUEL

2.1 The Beryllium Replacement Test Programme

Beryllium (Be) is currently used as the filler metal for induction brazing of appendages onto the sheaths of CANDU fuel elements (Figures 1 and 2). In CANDU fuel, all the metal components, including the appendages and the sheaths, are made of Zircaloy-4. While Be is an excellent filler metal for brazing of zircaloy, with compatible nuclear and corrosion properties for use in CANDU fuel, occupational exposure limits for Be have been reduced to very low levels because of its toxicity. In response to this change, the CANDU Owners Group initiated the Be Replacement Test Program with the objectives of identifying a filler material to replace Be and confirming that the brazed joints produced using the replacement material meet the established technical requirements for CANDU fuel.

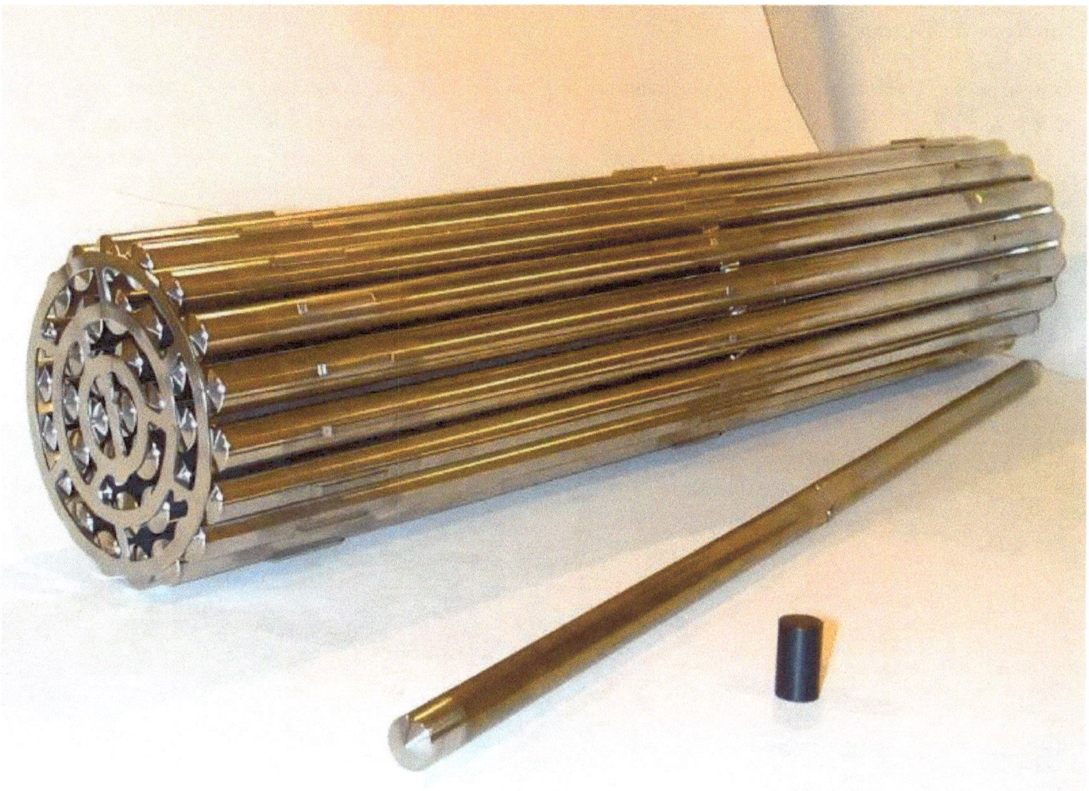

FIG. 1. Typical CANDU 37-element bundle, element, and UO_2 pellet.

FIG. 2. Close-up view of bearing pads and spacer pads on fuel elements, attached using induction brazing.

The Be Replacement Test Programme is comprised of a series of out-reactor tests, namely constructability assessment, mechanical property testing, corrosion testing, and high temperature transient testing, followed by an in-reactor test to demonstrate the suitability of potential candidate materials to replace Be. The candidate materials investigated in this study were stainless steel 316 (SS316), Inconel-600, Inconel-625, and Ni–2%Be alloy. The results from the multi-year testing programme, including observations from post-irradiation examination, have been reported in [5] and are highlighted in the following sections.

2.2 Out-Reactor Testing

One aspect of the brazing process that was looked at was the mechanical property of the brazed joint. The current technical specification states that "the brazed joint shall be strong enough to withstand forces on the appendage during fuel handling and in-reactor" [3]. Thus, the joint strength of test specimens manufactured using the candidate materials was considered acceptable when it was equal to or greater than the joint strength of the samples fabricated using Be as the braze filler material. Previous work has demonstrated that torque testing of brazed bearing pads represents a good measure of the joint's combined axial and circumferential shear strength [4]. The torque required to break the pad off the sheath is referred to as the torque strength. Results of torque tests conducted on joints brazed with Be and the different candidate materials indicate that the torque strengths of the candidate materials are comparable to Be. On average, the torque strengths of the brazed joints ranged from 120–143 N·m, regardless of the filler material used (Figure 3).

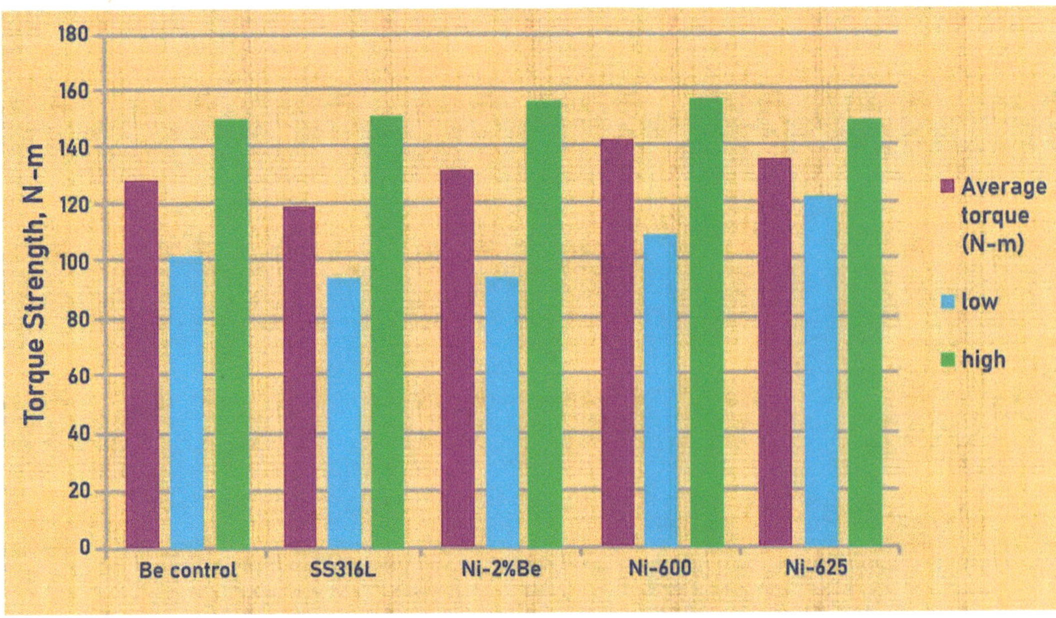

FIG. 3. Torque strength of joints brazed with different filler materials.

High temperature transient tests were performed to evaluate how the candidate materials behave under off-normal conditions, such as post-accident conditions (e.g. exposure to steam while at potentially elevated temperatures). The tests consisted of subjecting pressurized sections of sheaths with brazed appendages to continuous heating in air or steam environments. The behaviour of the candidate materials was compared to Be, mainly in terms of the sheath failure temperature. The specimens were tested in air or steam environments at a heating rate of 5°C/s or 25°C/s. All specimens were pressurized with He to 1.3 MPa, which was maintained constant during testing. Temperature and pressure readings were acquired during heating, and at the time the sheath failed, both temperature and pressure dropped rapidly. In general, the four candidate materials had failure temperatures similar to or higher than Be (Figure 4). On average, sheath failure temperatures ranged from 919°C to 961°C, with the candidate materials having slightly higher values than Be in every test condition. Therefore, all four alternative brazing materials were considered reasonable substitutes for Be based on the high-temperature transient tests.

Corrosion tests of samples brazed with various filler materials were conducted for 100 days under coolant conditions simulating normal reactor operating conditions: pressure of 10.4 MPa, temperature of 292°C, mass flux of 6000 kg/m^{2}s, Li levels of 0.6 mg/kg, average pH of 9.6, and dissolved oxygen level of 8 ppb. The specimens were weighed and visually inspected before and after the 100-day corrosion test. The weight change data showed that all specimens experienced a slight weight gain of 0.06 to 0.10 wt% of the specimen weights, with no significant differences among the filler materials. Also, visual inspection did not reveal any evidence of excessive corrosion or other abnormalities around the brazed joint. Thus, the candidate materials demonstrated similar or better corrosion properties relative to Be when exposed to coolant conditions representative of normal reactor operations.

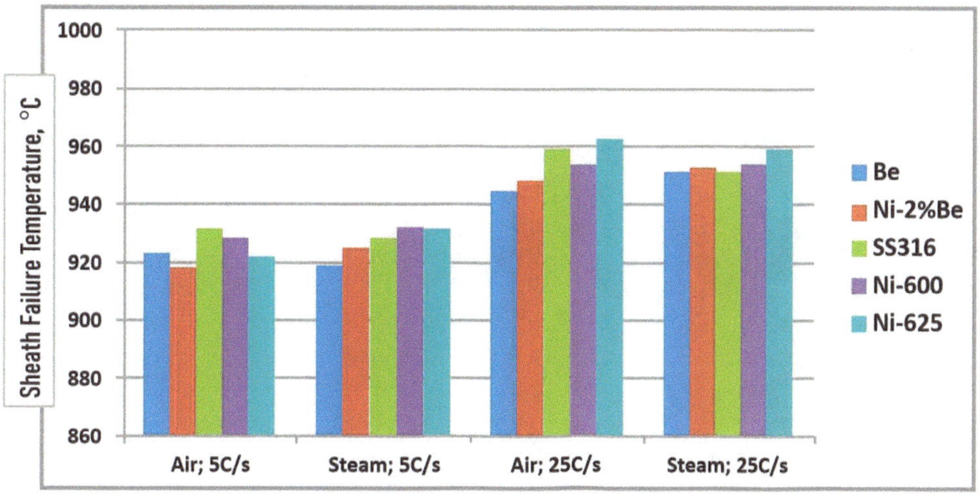

FIG. 4. Sheath failure temperatures of specimens brazed with the different candidate materials under the various test conditions.

The specimens' torque strength and hydrogen contents were determined after the corrosion test. The hydrogen contents were measured using hot vacuum extraction mass spectrometry. Results showed that the Ni-based materials picked up significant amounts of hydrogen. As listed in Table 1, the hydrogen contents of the specimens with Ni-containing alloys were 136–155 ppm, compared to typical as-fabricated values of 10–20 ppm, a significant increase of about ten times. In contrast, the Be and SS316 brazed specimens only had hydrogen contents of 16–21 ppm, an insignificant uptake of hydrogen from the corrosion test. The significant hydrogen uptake of Ni-containing alloys appeared to have affected the torque strength. As shown in Table 1, the torque strengths of the Ni-based alloys decreased from the as-fabricated values of 120–140 N·m to the post-corrosion values of 83–105 N·m. The torque strengths of Be and SS316 specimens were unaffected by the corrosion test, i.e. as fabricated values of 128 and 120 N·m, respectively, compared to post-corrosion values of 136 and 116 N·m.

Given the confirmed deleterious effect of Ni on the hydrogen uptake of Zircaloy as shown by the degraded torque strength and elevated hydrogen content of the Ni-containing specimens, it was decided that the only candidate material going into the irradiation testing phase was SS316.

TABLE 1. HYDROGEN CONTENT AND TORQUE STRENGTH OF POST-CORROSION TEST SAMPLES

Braze Material	Hydrogen Content, ppm	As-fabricated Torque Strength, N-m	Post-Corrosion Torque Strength, N-m
Beryllium	16	128	132
SS316	21	120	116
Inconel-600	136	142	83
Inconel-625	139	136	105
Ni-2%Be	155	132	92

2.3 In-Reactor Testing

The in-reactor test involved irradiation of demountable fuel elements (DMEs) at the National Research Universal reactor test loops with the fuel elements containing either Be as braze filler material (as the control) or SS316 as the candidate material. The DMEs were assembled to ensure that for each DME with Be-brazed appendages, there were two DMEs with SS-brazed appendages on either side (Figure 5). The DMEs were irradiated for a total of 150 full power days of reactor operation in the National Research Universal. As shown in Figure 6, the DMEs experienced burnups of around 180 MWh/kgHE, and average linear powers of around 51 kW/m at the beginning of life, generally declining to around 47 kW/m at discharge.

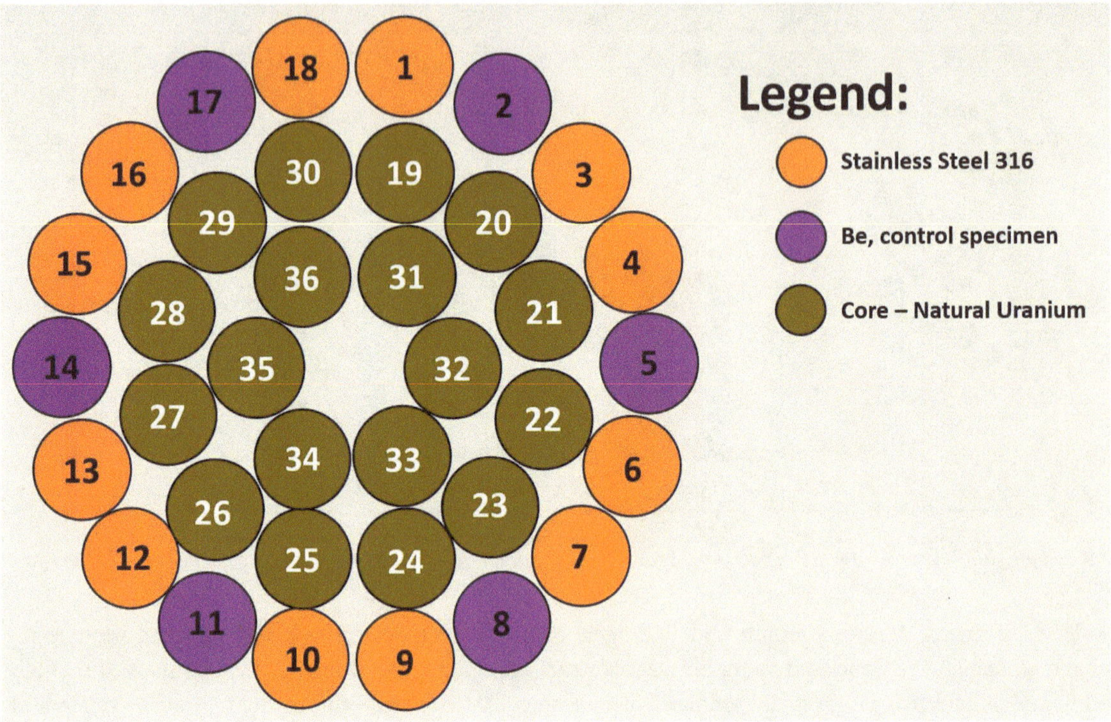

FIG. 5. Schematic of fuel bundle cross-section for the irradiation test of Be and SS316-brazed DMEs.

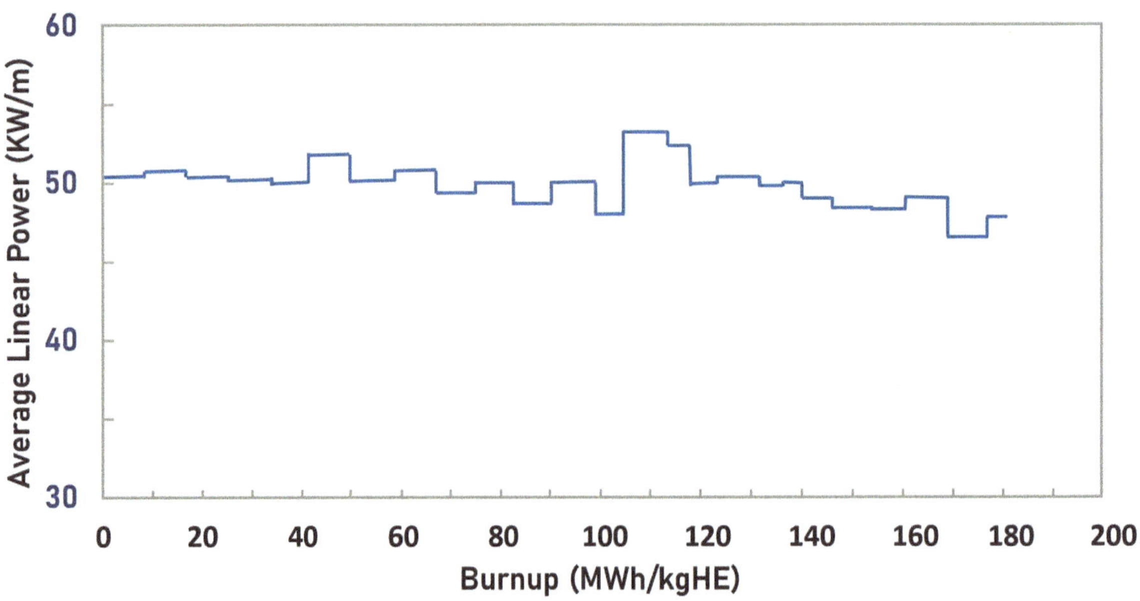

FIG. 6. Power history of Be and SS316-brazed DMEs.

2.4 Post Irradiation Examination

Upon visual examination using the in-cell stereomicroscope, all elements appeared to be intact and in good condition. No anomalies were observed on the sheath, and typical, minor wear was observed on the appendages. Specimens were cut from several DMEs for hydrogen content analysis using hot vacuum extraction mass spectrometry. As shown in Table 2, hydrogen content in the SS316 samples were almost double that of the Be braze elements, 103 ppm H vs 54 ppm H, respectively. Four DMEs were subjected to metallographic examination: two Be control elements and two SS316 braze elements. Based on a detailed examination, the hydride distribution was found to be essentially the same between the Be-brazed and SS-brazed joints (Figure 7).

TABLE 2. HYDROGEN AND DEUTERIUM CONTENTS OF POST-IRRADIATION TEST SAMPLES

Element	Braze	H Content (ppm)	D Content (ppm)
BL03	SS316	105 ± 7	0.4 ± 0.1
BL05	Be	55 ± 5	<0.10
BL06	SS316	95 ± 6	<0.10
BL09	SS316	96 ± 5	<0.10
BL11	Be	58 ± 6	<0.20
BL12	SS316	110 ± 7	<0.20
BL15	SS316	113 ± 7	<0.20
BL17	Be	49 ± 5	<0.20
BL18	SS316	96 ± 5	<0.10
Be Braze Average		54 ± 5	<0.2
SS316 Braze Average		103 ± 8	<0.2

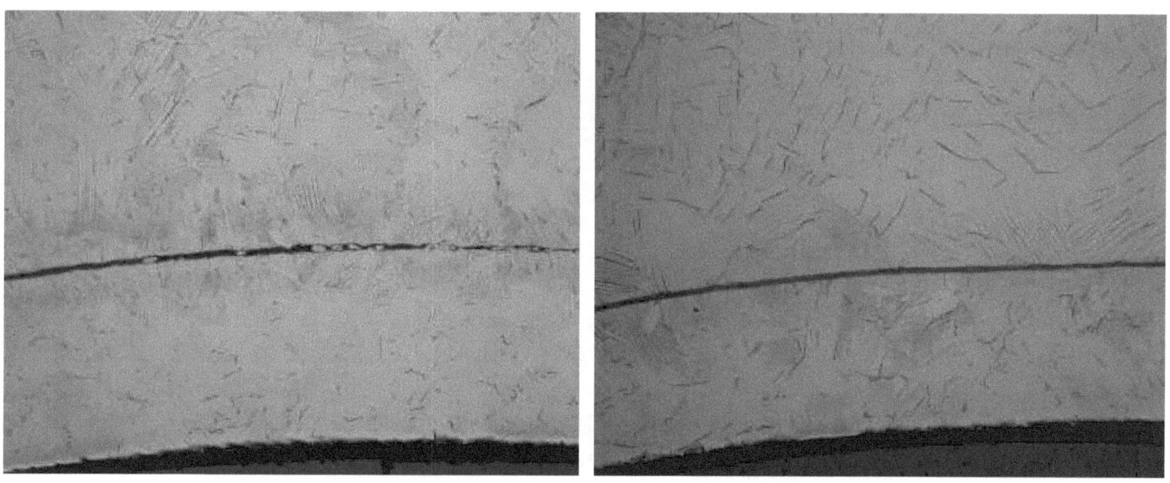

Be Braze Element – BL17 SS316 Braze Element – BL18

FIG. 7. Typical Hydride Distribution for Be and SS316-brazed DMEs.

Two irradiated samples (one Be brazed, one SS316 brazed) were subjected to Vickers hardness testing, and the location and results are shown in Table 3. To assess the effect of irradiation, unirradiated samples were also hardness tested at the same locations. The sheath hardness and bearing pad hardness values were comparable between the two different brazed elements, well within the standard deviation of the measurements. Vickers hardness values for the brazed areas differed, with the SS316 braze being slightly harder than the Be braze. In the unirradiated condition, the SS316 Vickers hardness values were about 15% harder than Be, while in the irradiated condition, the SS316 Vickers hardness values were about 11% harder than Be. Overall, the Be and SS316 Vickers hardness values were comparable, and both showed a similar extent of irradiation hardening.

TABLE 3. VICKERS HARDNESS TESTING RESULTS FOR THE IRRADIATED DMES

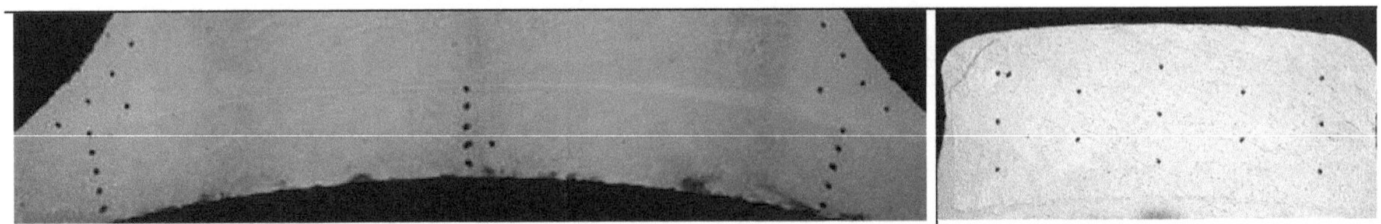

Sheath Condition	Braze Hardness*		Sheath Hardness*		Bearing Pad Hardness*	
	Be	SS316	Be	SS316	Be	SS316
Unirradiated	270 ± 17	311 ± 19	207 ± 25	206 ± 26	193 ± 12	182 ± 16
Irradiated	327 ± 8	363 ± 23	239 ± 15	236 ± 16	269 ± 15	280 ± 23

*Hardness values are Vickers Hardness using the 100g indenter (HV 0.1); values given as average ± 1 std deviation.

Torque strength measurements were performed on 18 brazed joints – 12 SS316 and 6 Be. The measured torque strength values were comparable between the two braze materials, with the Be brazed appendages being slightly stronger on average (112 N·m vs. 104 N·m, see Table 4). However, based on the standard deviation of the measurements, this slight difference is not statistically significant.

TABLE 4. TORQUE TESTING RESULTS FOR THE IRRADIATED DMES

Parameter	Bearing Pad Torque Strength Measurements (N·m)	
	Be Braze	**SS316 Braze**
Average	112	104
Standard Deviation	5	12
Maximum	120	120
Minimum	105	85

2.5 Conclusions of the Be Replacement Programme

Employing a series of out-reactor tests, including mechanical property testing, corrosion testing, and high temperature transient testing, several potential candidate materials were investigated to assess their suitability as replacements to Be as a filler material in brazing of appendages for CANDU fuel. Results of the out-reactor testing phase showed that among the candidate materials, SS316 demonstrated the most likelihood of success. Several DMEs containing both SS316 and Be as braze filler materials were manufactured and irradiation tested in the National Research Universal test loops for 150 full power days under typical operating conditions. Post-irradiation examination of the irradiated DMEs demonstrated that the SS316-brazed element's performance was similar to that of the industry standard Be-brazed elements. Different rate of hydrogen pickup was the only major difference observed, however, this observation is believed to have minimal impact on the fuel's performance given the similar hydride distribution and mechanical properties of the brazed joints.

3. ADVANCED CLADDING ALLOYS

3.1 Zr-Nb Alloys

The introduction of niobium-containing zirconium cladding alloys, specifically ZIRLO (Zr-1 wt% Sn-1 wt% Nb) and M5 (Zr-1 wt% Nb), has led to the replacement of Zircaloy-4 in several PWRs [6]. One distinctive advantage of this family of alloys compared with Zircaloys is their improved corrosion resistance during normal operating conditions, especially at higher burnup levels and higher coolant temperatures [6-8].

The objective of this work is to assess the feasibility of using Zr-Nb alloy for fabricating fuel sheath in CANDU reactors. Part of the work is to assess the potential undermining effect of a Heat Affected Zone produced during CANDU fuel manufacturing (i.e. brazing and welding) on the mechanical properties of Zr-Nb alloys. The assessment was based on comparing tensile properties of as-received and brazed Zr-Nb materials.

The analysis of the chemical composition of the Zr-Nb alloy was conducted using inductively coupled plasma mass spectrometry, and the corresponding results are presented in Table 5. The as-received tubes have the following dimensions: outer diameter of around 9 mm and inside diameter of 8 mm.

TABLE 5. ELEMENTAL COMPOSITION OF ZR-NB ALLOY

Element	Concentration in Tube (μg/g)	Element	Concentration in Tube (μg/g)
Beryllium	<1.4	Molybdenum	1.3 ± 0.3
Boron	<60	Nickel	22 ± 3
Cadmium	<4	Niobium	10000 ± 1000
Chromium	64 ± 7	Sodium	<40
Cobalt	<1.7	Tantalum	7.1 ± 0.7
Copper	10 ± 1	Tin	<5
Hafnium	57 ± 6	Titanium	<19
Iron	510 ± 60	Tungsten	7.9 ± 0.8
Lead	<1.1	Uranium	<4
Lithium	<6	Vanadium	<3
Manganese	4.3 ± 1	Zirconium	Balance
Magnesium	<20		

3.2 Effect of Brazing

To determine the suitability of current CANDU fuel manufacturing techniques for Zr–Nb alloy, it was necessary to study the effect of joining techniques, such as brazing and welding, on the resulting material properties of this alloy. Subsequently, several samples were subjected to a simulated brazing operation prior to testing. Brazing was simulated by heating up the specimen to about 1160°C in approximately 100 seconds under an inert atmosphere followed by cooling to 700°C in approximately 20 seconds.

3.3 Tensile Testing

Tensile testing was done in accordance with the ASTM standard E8M-15a [9] and ASTM E21-03 [10] in order to obtain experimental data for the 0.2% yield strength/stress, ultimate tensile strength, and total elongation of specimens in both as-received and heat-treated metallurgical conditions. The geometry and dimension of tensile test samples are schematically illustrated in Figure 8, and the configuration for the tensile test is shown in Figure 9. Tensile properties of the tube specimens were determined by uniaxial tensile tests in the axial direction. For each test condition, three specimens were tested. The material property measurements were then averaged from these three runs to ensure consistency.

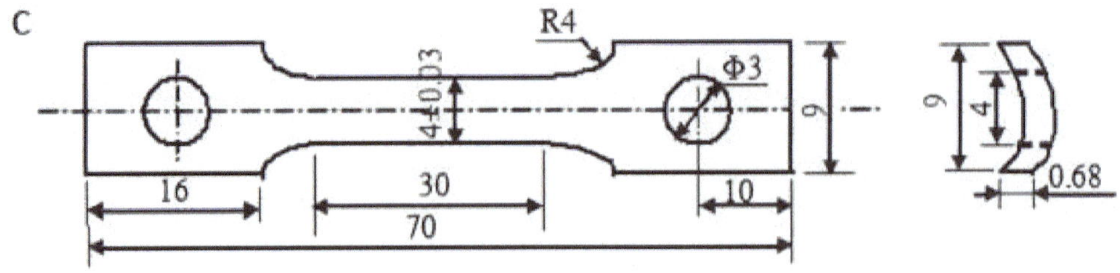

FIG. 8. Schematic of tensile test specimen.

Tensile tests were performed in air at different temperatures, namely 25°C, 100°C, 275°C, and 375°C. In accordance with ASTM E21-03 [10], the specimens were subjected to a soak time of 30 minutes at test temperature prior to testing to ensure that the temperature was uniform throughout the material. Two thermocouples were attached to the top and bottom of the sample gauge length respectively to measure the temperature (see Figure 9). The reported temperature was the average of the top and the bottom readings for each test. As specified in the ASTM standards, the specimens were tested using a strain rate of 5.0×10^{-4}/s.

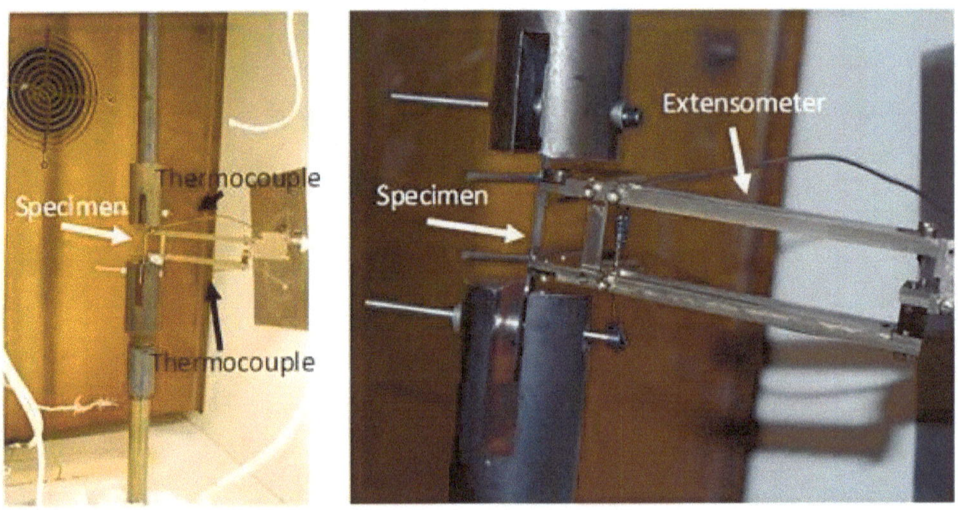

FIG. 9. Photos showing tensile test configuration.

3.4 Tensile Test Results

Tensile test results are presented as stress – strain curves, one for each of the different test temperatures (25°C, 100°C, 275°C, and 375°C) in Figure 10 for both as-received and heat-treated samples. From these data, the tensile properties of Zr–Nb tubes as affected by temperature are re-plotted in Figure 11. Both yield strength and ultimate tensile strength decreased as the testing temperature increased, regardless of the material's metallurgical condition. The ductility of as-received Zr–Nb tubes slightly increased with the test temperature, but it showed no significant change for the heat-treated samples. Heat treatment brought about only slight changes in yield strength and ultimate tensile strength of Zr–Nb tube specimens at all test temperatures. However, the brazing heat treatment led to a pronounced decrease in ductility at all test temperatures (Figure 11).

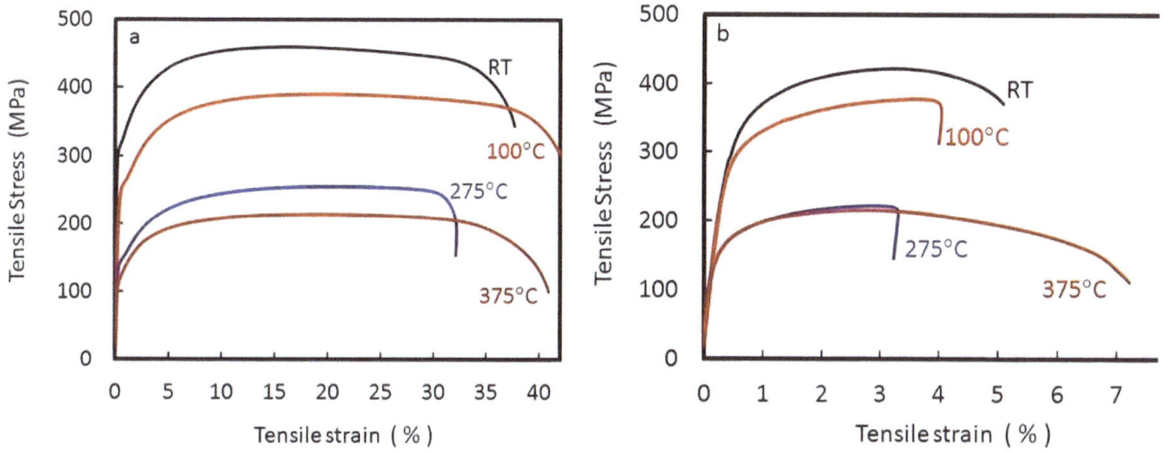

FIG. 10. Typical stress–strain curves obtained from tensile testing of Zr–Nb tube specimens at different conditions (a) as received and (b) Heat Affected Zone.

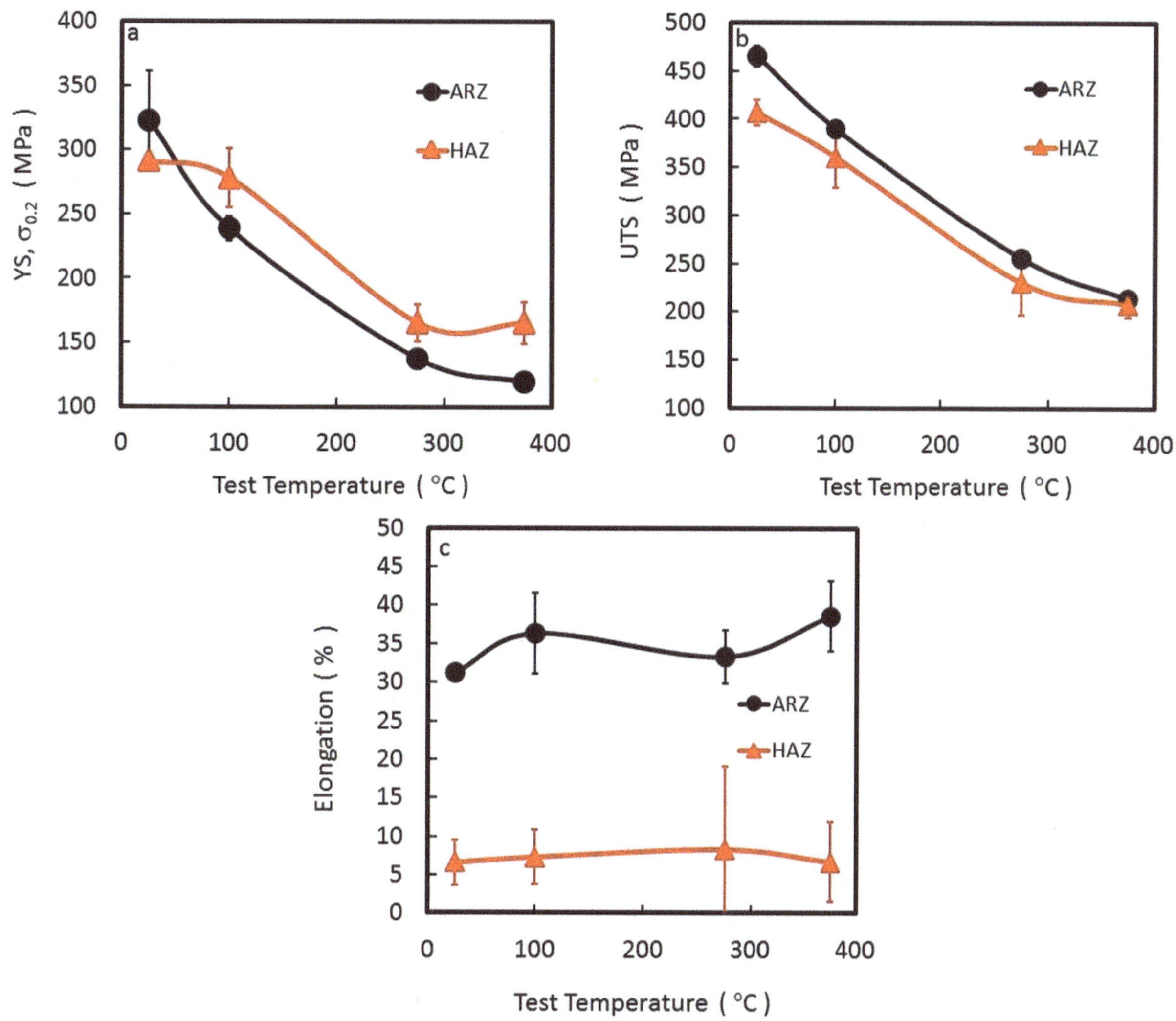

FIG. 11. Effect of heat treatment on (a) yield strength, (b) ultimate tensile strength, and (c) ductility of Zr–Nb tube samples at different temperatures.

3.5 Conclusions To-Date of the Advanced Cladding Programme

To determine the suitability of current joining technology used for present day CANDU fuels to advanced zirconium alloys, samples of Zr–Nb alloy were tested in their as-received condition, as well as in the heat-treated condition simulating the brazing process typically employed for attaching appendages to CANDU fuel sheath. Specimens were tested for their mechanical properties through tensile tests to obtain data for yield strength, ultimate tensile strength, and total elongation. These tests were conducted in air at different temperatures ranging from room temperature to 375°C.

The results demonstrate that the heat treatment has an observable effect on the tensile properties of Zr–Nb alloy. While the relative degradation of yield strength and ultimate tensile strength due to the simulated brazing operation was not significant (around 10% decrease), there was a significant reduction in ductility by more than three-fold, i.e. from around 35% elongation for the as-received samples to less than 10% for the heat-treated ones.

4. SUMMARY AND FUTURE WORK

A sampling of the development work on fuel fabrication technologies being conducted at CNL was discussed using two specific examples: one is a study of new filler materials for brazing of zirconium alloys, and another is the work on advanced fuel cladding alloys. Results of the out-reactor and in-reactor tests demonstrated that the SS316-brazed element's performance was similar to that of the industry standard Be-brazed elements. Different rate of hydrogen pickup in-reactor was the only major difference observed, however, this observation is believed to have minimal impact on the fuel's performance given the similar hydride distribution and mechanical properties of the brazed joints. The next steps for this work may include demonstrating this material's performance in an actual operating CANDU reactor but this decision will be made by the respective utilities.

The work on Zr–Nb alloy was discussed in terms of tensile test results of as-received and heat treated (to simulate brazed condition) samples. One important consideration for this study was to determine the suitability of current CANDU fuel manufacturing techniques for Zr–Nb alloy, particularly in terms of the effect of joining techniques, such as brazing, on the resulting material properties of this alloy. Results indicate that while brazing does not have a significant impact on yield strength and ultimate tensile strength of Zr–Nb alloy, a significant reduction in ductility was observed. The reduction in ductility may very likely affect impact strength of the material, which is important to ensure the integrity of fuel bundle during transportation and handling. It is not clear whether this reduction in ductility will translate into a serious performance issue, especially when the effect of irradiation embrittlement is taken into consideration. This possible issue needs to be explored in any future investigation.

Additionally, it is recommended that further study is conducted to either improve current joining techniques of brazing and welding to make them more suitable for Zr–Nb alloys or develop alternative joining techniques that will result in minimizing the effects of heat treatment on this alloy.

ACKNOWLEDGEMENT

Work by CNL presented in this paper was conducted under the auspices of the Federal Nuclear Science and Technology Work Plan funded by Atomic Energy of Canada Ltd. and the Safety and Licensing R&D Program funded by the CANDU Owners Group.

REFERENCES

[1] AECL, Federal Nuclear Science & Technology Work Plan – AECL (March 2024) https://www.aecl.ca/science-technology/federal-science-and-technology-work-plan.
[2] CANDU OWNERS GROUP, Pages - Research & Development (March 2024) https://www.candu.org/Pages/RD.aspx.
[3] AECL TECHNICAL SPECIFICATION, Beryllium-Brazed Appendage Joints on Reactor Fuel Sheathing, TS XX-37357-1, Revision 1 (2000).
[4] HARMSEN, J., et al., Beryllium Brazing Considerations in CANDU Fuel Manufacture, Paper presented at the 11th International CANDU Fuel Conference, Niagara Falls, Canada (2010).
[5] DIMAYUGA., F., et. al., Stainless Steel as a Replacement for Beryllium as Braze Material, Paper presented at the 15th International Conference on CANDU Fuel, Ajax, Ontario, Canada (2022).

[6] COX., B., et al., Fuel Material Technology Report V.1, Advanced Nuclear Technology International, Skultuna, Sweden (2006).
[7] YEGOROVA, L., et al., Experimental Study of Embrittlement of Zr–1%Nb VVER Cladding under LOCA-Relevant Conditions; REP/NUREG-IA-0211/2005; IRSN 2005-194; NSI RRC KI 3188 (2005).
[8] SHELBALDOV, P. V., et al., E110 Alloy Cladding Tube Properties and Their Interrelation with Alloy Structure-Phase Condition and Impurity Content, Zirconium in the Nuclear Industry, Twelfth International Symposium, ASTM STP 1354 (2000) 545-559.
[9] AMERICAN SOCIETY FOR TESTING OF MATERIALS, Standard Test Methods for Tension Testing of Metallic Materials, ASTM E8/E8M-15a (2015).
[10] AMERICAN SOCIETY FOR TESTING OF MATERIALS, Standard Test Methods for Elevated Temperature Tension Testing of Metallic Materials [Metric], ASTM E21-03 (2003).

ID#13
INDUSTRIAL DEVELOPMENT OF N36 ZIRCONIUM ALLOY CLADDING TUBES

C. XU, W. ZHAO, Z. YANG, X. DAI, J. QIU
Science Technology on Reactor Fuel and Materials Laboratory,
Nuclear Power Institute of China (NPIC),
Chengdu, China

Abstract

The development of nuclear power technology is closely related to the improvement of FA cladding materials. Zirconium alloy is currently the only cladding material used in water-cooled nuclear reactors. In order to meet the requirements of high burnup and long service life of FAs in nuclear power technology, the NPIC has been conducting research on high-performance zirconium alloys since the 1990s. Through composition design and optimization experiments, microstructure research, out-of-pile corrosion, and mechanical performance tests, a high-performance zirconium alloy with comprehensive properties superior to Zr-4 alloy was invented and named N36.

In the industrial preparation process of N36 zirconium alloy cladding tubes, the ingot scale was continuously enlarged from 260 kg, 320 kg, 500 kg, 1000 kg, and 3000 kg to 5000 kg. The research was conducted at various stages, including process exploration, process verification, process scaling, process optimization, process standardization, and industrialization, to establish the manufacturing processing route and technology for N36 zirconium alloy cladding tubes. The prepared N36 zirconium alloy cladding tubes meet the design technical requirements for corrosion resistance, hydrogen absorption, tensile strength, burst strength, creep, fatigue, texture, and corrosion stress rupture. The microstructure is uniform, with grain size above level 11(~3–4 μm), and the second-phase particles are finely dispersed with an average size of approximately 80 nm. The N36 zirconium alloy cladding tubes exhibit significantly better performance in terms of out-of-pile corrosion resistance, hydrogen absorption resistance, creep resistance, fatigue resistance, iodine-induced stress corrosion resistance, and high-temperature oxidation resistance compared to Zr-4 alloy.

Furthermore, characteristic fuel rods and leading FAs were prepared using N36 zirconium alloy cladding tubes, and they successfully underwent irradiation tests in commercial reactors for 60 months and 54 months. The fuel rod burnup reached 56 GWd/tU and 55 GWd/tU, respectively. The poolside inspection results showed that the fuel rod structure was intact, and the surface oxide film was dense and smooth. The N36 zirconium alloy exhibited good resistance to irradiation corrosion, hydrogen absorption, creep, and growth. The post-irradiation examination (pool side inspection and hot cell examination) of the fuel rods further confirmed the excellent in-reactor performance of N36 zirconium alloy. Through this research, comprehensive knowledge of the industrial manufacturing process and application technology of N36 zirconium alloy cladding materials were achieved, and data on the in-pile and out-of-pile performance of N36 zirconium alloy were obtained, thereby realizing the goal of engineering application.

At present, the manufacturing technology of N36 zirconium alloy cladding tubes has been successfully transferred to Western Energy Material Technologies Co. Ltd, marking the first industrial scale production of advanced nuclear-grade zirconium alloy materials in China. As the CF3 FAs cladding material for China's third-generation NPP, “Hualong One”, N36 zirconium alloy has already been used for the replacement of the first batch of FAs in the K2 and K3 nuclear reactors in Pakistan. With the large scale development of nuclear power in China and the implementation of the strategy of "going global" with nuclear power, N36 zirconium alloy has broad prospects for application.

ID#14

DEVELOPMENT STATUS OF ATF AND 3D PRINTING TECHNOLOGY DEVELOPED BY KAERI FOR NUCLEAR APPLICATION

H.G. KIM
KAERI,
Daejeon, Republic of Korea

Abstract

This paper deals with new technologies for manufacturing nuclear fuel cladding and reactor core components. The first is a surface treatment technology for short-term ATF cladding to reduce the hydrogen explosion problem that occurred in the Fukushima accident, and the second is a 3D printing (AM) technology that allows the manufacturing of complex shaped parts and restoring damaged parts. KAERI's research conducted since 2013 described safety, economic feasibility, and various technical considerations necessary for the commercialization of ATF cladding tubes incorporating short-term coating methods. In addition, we covered the matters necessary to manufacture and apply new materials and parts for NPPs by applying 3D printing technology, which has recently become an issue in the nuclear industry.

1. INTRODUCTION

As efforts to reduce carbon emissions have recently progressed around the world, demand for nuclear power generation with almost no carbon emissions is increasing. In addition, much attention is being focused on the safe operation of existing NPPs and the development of small nuclear reactors with improved safety. In this situation, research on ATF to improve accident safety of nuclear fuel in water-cooled NPPs is reaching the commercialization stage Another issue is that efforts are being made to introduce 3D printing as an innovative manufacturing technology in the repair of damaged parts of large operating NPPs, and the production of discontinued parts or the development of key parts for new small reactors.

In this context, the first topic is the development of ATF cladding, and the second topic is the 3D printing (AM) technology for application to nuclear industry. Research to improve the safety and economic feasibility of operating NPPs has been ongoing for a long time. In particular, after the Fukushima accident in 2011, research to develop ATF has been conducted [1-6], and it is also mentioned in the EU taxonomy. Another change in the nuclear industry is that innovative technologies such as 3D printing technology are gradually being applied to the conservative nuclear industry [7-11].

The development of ATF has been conducted for more than 10 years in the NPP operating country. Regarding the ATF cladding development, various studies have been performed to improve oxidation resistance by using coating technology, Fe alloy and SiC material application, as well as, to improve deformation resistance at high temperature by using oxide dispersion strengthening technology, Mo or SiC material application. As research progresses, coating technology is being established as a target for short-term commercialization in consideration of economic feasibility, compatibility, and application time. The KAERI has developed various coating technologies using AIP and 3D printing to improve high-temperature oxidation resistance for the Cr or CrAl, FeCrAl, etc. and oxide dispersion strengthened (ODS) technology using 3D printing for Y_2O_3 particles to improve high-temperature deformation resistance with the concept of ATF cladding However, there are requirements to consider for rapid commercialization of short term oxidation-resistant coating technology. Mass productivity is a very important economic issue, and if a change in the base material of the zirconium clad tube occurs during coating, activities to develop a new zirconium alloy will be added compared to the advantage of using the existing commercialized alloy. Therefore, it is important to develop oxidation-resistant coating technology that allows for mass production and does not affect commercialized zirconium alloys. This paper not only presents various research details of KAERI related to ATF cladding but also presents AIP coating technology that can be mass-produced without changing the zirconium alloy base material for rapid commercial application.

Internationally, research is underway to manufacture parts using 3D printing technology and apply them to commercial reactors and ultra-small nuclear reactors [7-11]. 3D printing is being developed to manufacture material parts for NPPs. These technologies are accelerating throughout the industry with the development of raw materials for 3D printing, printer and control technology, additive process and analysis technology, and QC related technology. As 3D printing technology is gradually expanded and applied to the nuclear field, a systematic approach to technology development purpose, direction, requirements, commercialization, and quality verification is required. KAERI is also developing core technologies for producing nuclear fuel, core components, reactor vessels, and sensing modules. The developed technology could be used as a cross-cut technology that can be commonly used in various reactor types. Therefore, it is necessary to introduce the systematic strategy and necessary technical specifications required for 3D printing technology development.

ATF technology is expected to be very useful in improving the safety of operating NPPs and water-cooled SMR NPPs, and in high burn-up operations to improve economic efficiency in the future. 3D printing technology is expected to be widely used in the maintenance of operating NPPs and the manufacturing of innovative parts for new SMR NPPs. This paper describes KAERI's development content and development direction for the two core technologies mentioned above that have recently received attention in the nuclear power industry.

2. DEVELOPMENT STATUS OF ATF AND 3D PRINTING TECHNOLOGY AT KAERI

2.1 Development Status of ATF Cladding for LWRs

2.1.1 Considering the ATF concept for near-term applications

Since 2013, KAERI has developed a coating technology to improve high-temperature oxidation resistance and ODS technology to improve high-temperature deformation resistance with the concept of ATF fuel cladding as shown in Table 1. In this way, combining the coating and ODS technology to improve the performance of the zirconium alloy surface can be defined as a surface modification technology.

TABLE 1. ATF CLADDING CONCEPTS DEVELOPED BY KAERI FOR NEAR-TERM APPLICATION

No	Partial ODS		Coating		Remarks (target)
	Oxide type	method	material	method	
A-1	-	-	Cr, CrAl	AIP/Cold spray/3D printing	Corrosion
A-2	Y_2O_3	Laser beam scanning	Cr, CrAl	AIP/Cold spray	Strength and Corrosion
B-1	-	-	Cr/FeCrAl/SiC	AIP for Cr, 3D laser coating/Cold spray for FeCrAl, SiC	Corrosion
B-2	Y_2O_3	Laser beam scanning	Cr/FeCrAl	AIP for Cr, 3D laser coating/Cold spray for FeCrAl	Strength and Corrosion

KAERI's short term ATF cladding development technology gradually improved materials and coating technology by applying nuclear fuel rod design, prototype manufacturing, performance evaluation, and quality confirmation methods according to materials and various coating processes, as shown in Figure 1. In addition, we attempted to improve the completeness of the results by continuously improving the performance of materials and manufacturing processes by reflecting the evaluation results.

The reason for developing oxidation-resistant material coating technology on existing zirconium alloy with short-term ATF cladding technology is the result of prioritizing ease of development and economics. Using existing zirconium alloys has the advantage of reducing the cost and time required for new materials or new large scale facilities to manufacture them. However, it was confirmed that even if it is a short term technology, there may be differences in bonding strength or changes in the base material of the zirconium cladding tube depending on the coating method.

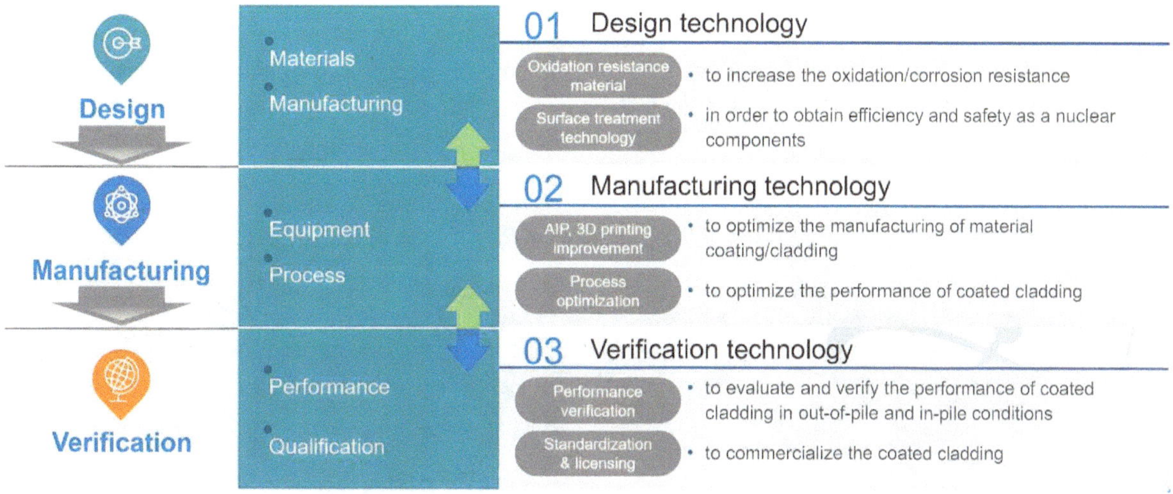

FIG. 1. Accident Tolerant Fuel (ATF) Cladding Development Process at KAERI.

Regarding coating materials, most coating cladding concepts are currently prioritizing Cr coating, but KAERI has been evaluating the applicability of not only Cr but also potential materials such as Si, CrAl, SiC, and FeCrAl alloys since 2013 [2, 3, 7, 12, 13]. Coating materials have excellent high-temperature oxidation resistance due to their inherent physical properties, but their properties vary depending on temperature, pressure, and water-chemical environment, such as dissolution in normal operating environments, so accurate performance verification of the material is necessary.

Therefore, to successfully develop ATF cladding technology, it is necessary to understand the contents of existing zirconium alloy cladding tubes, including manufacturing, required performance under normal conditions, performance requirements under accident conditions, as well as spent nuclear fuel management as shown in Figure 2. Based on the performance evaluation of commercial Zr cladding tubes, surface modification was introduced as a technology to improve performance in an accident environment when developing ATF cladding tubes.

Performance parameters		Commercial Zr alloy cladding
Manufacturability		• Optimized
Normal condition	Corrosion	• No problem **(Oxide Thickness <100mm)**
	Creep	• No problem **(Hoop Strain<1%)**
	Irradiation growth	• No problem **(Shoulder gap closure)**
	Wear resistance	• No problem **(Wear damage < 10% in thickness)**
	PCMI/PCI	• PCMI/PCI failure related to the hydride (hydrogen pick-up by corrosion)
DBA	LOCA	· Severe oxidation and hydrogen generation · Ballooning & Burst · Limitation **(PCT<1204°C, ECR<17%)**
	RIA	· PCMI failure related to the hydride
BDBA		· Severe oxidation and hydrogen generation · Melting problem
Spent fuel storage	Failure mechanism	• Brittle behavior by hydride • Hydride re-orientation problem **(Fuel-clad interface temp.: drying: <570°C, storage: <400°C) (Clad hoop stress: <90MPa)**

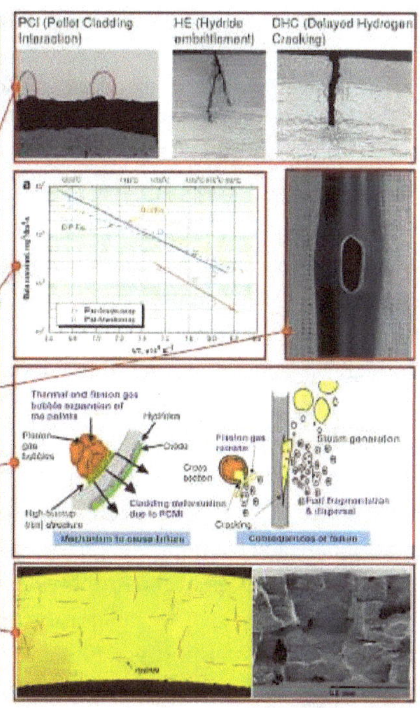

FIG. 2. Performance Issues related to Manufacturing, Normal Operating, Accident Conditions, and Storage Environment of Commercial Zr Alloy Cladding Tubes.

In detail, technology development classified each coating material into whether it could be coated with a plasma spray, cold spray, AIP, and 3D printing, and researched and developed each coating condition as shown in Figure 3. Research and development concern the suitability of various coating technologies, taking into account the physical properties of the coating material, the homogeneity of the coating film, adhesion, and compatibility with the base material.

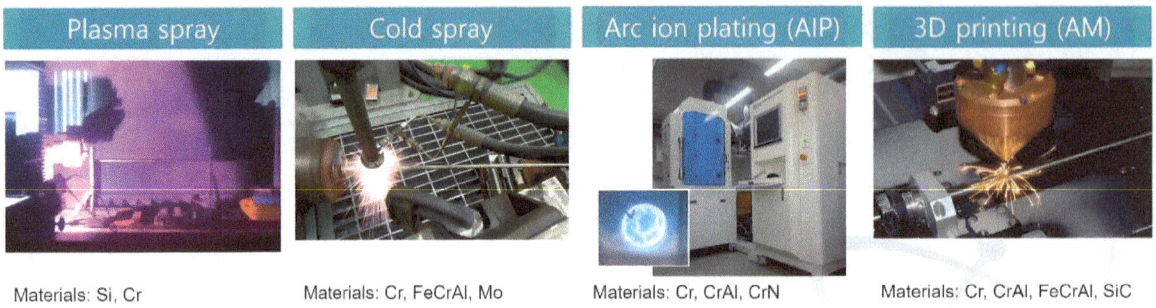

FIG. 3. Applied Coating Methods and Materials for Accident Tolerant Fuel (ATF) Cladding Development at KAERI.

The main purpose of the ATF cladding development technology based on oxidation-resistant material coating is to improve safety margins and reduce hydrogen generation by improving oxidation resistance in normal and accident environments. Therefore, although it is possible to secure improvement in oxidation resistance with coating technology alone, it does not play a significant role in reducing the amount of plastic deformation and rupture of the cladding tube during an accident. To solve this problem, KAERI proposed and developed a 3D printing technology to improve high-temperature strength by partially dispersing fine oxides into the zirconium alloy base material [8].

In this way, KAERI is developing various surface treatment technologies that improve oxidation resistance and high-temperature deformation resistance while being advantageous for economic efficiency and commercialization by using commercial zirconium alloy base materials as a short-term technology after the Fukushima accident.

2.2.1 Research results of surface modified-ATF cladding developed at KAERI for near-term applications

While developing ATF cladding surface treatment technology, KAERI has studied each coating technology and combination of coating materials to identify the advantages and disadvantages and develop know-how technology to solve problems. Considerations for coating materials and methods applied to nuclear zirconium alloy cladding tubes have been summarized in the ASTM paper [2]. Below is a description of the research and experience for applying coating methods to zirconium base materials.

The experience gained from coating oxidation-resistant materials with plasma spray had the advantage of fast coating speed but had the problem that the density of the coating layer was low, and that the zirconium alloy base material recrystallized when the temperature was increased under the condition of increasing density. One way to increase density and improve interface properties is to additionally apply a re-dissolution method using a laser heat source [2]. Additionally, research has shown that Si coatings have the problem of easily dissolving under normal corrosion conditions [7].

Cold spray coating has the advantage of having a faster coating speed and a higher density coating layer than plasma spray and has excellent adhesion, so it has high potential as a coating technology [14]. However, there is a need to understand the impact of the heterogeneous interface formed between the coating layer and the zirconium base material in the accident environment, and the surface roughness after coating is large, so follow-up treatment such as polishing is essential. In addition, it is to be considered that additional costs are incurred when using He rather than Ar as a transport gas to supply powder during coating. If oxidation occurs on the bonding surface between particles as well as diffusion of oxygen along the bonding surface between particles during long-term exposure under normal corrosion conditions due to mechanical bonding caused by high speed collision of particles during coating, the problem of delamination needs to be also verified.

Coating using AIP is a technology that can obtain a homogeneous coating film, design a large device, enable mass production per batch, and minimize damage to the base material. KAERI developed Cr coating technology using AIP on zirconium base material [3]. Cr coating using AIP has the advantage of excellent high-temperature oxidation resistance, and excellent performance has been confirmed by developing CrAl alloy, which has better oxidation resistance than Cr [12-14]. As a result of the thermo-gravimetric analyser test, the Cr-coated zirconium alloy using AIP was confirmed to have a sufficient role in preventing oxidation, as shown in Figure 4, and to maintain its soundness without peeling in a high temperature steam environment of 1200°C (Figure 5).

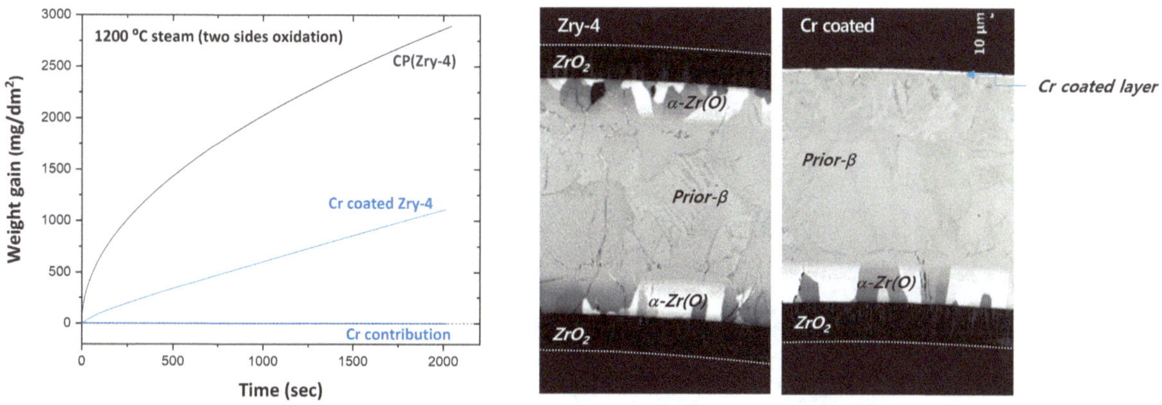

FIG. 4. Evaluation of High-temperature Oxidation Characteristics of Un-coated Zr Cladding and Cr-coated Zr Cladding using AIP Coating Method and Cross-sectional Observation after Oxidation Test.

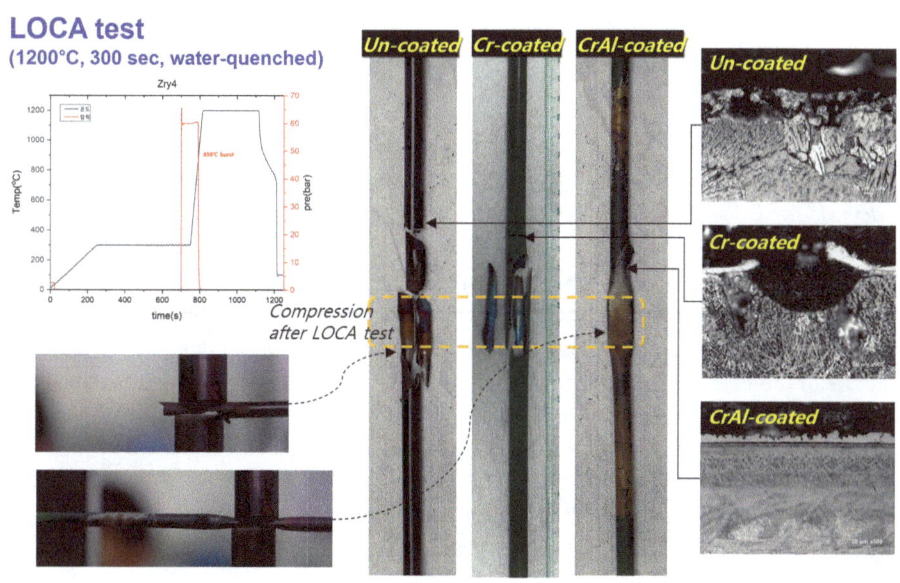

FIG. 5. Damaged Cladding Shape and Cross-sectional Microstructure of the Un-coated Cladding, Cr-coated Cladding, and CrAl-coated Cladding by AIP Coating Method after LOCA Simulation Test and then Compression Test.

While developing and evaluating Cr coating technology using AIP, KAERI developed CrAl alloy to reduce the difference in thermal expansion coefficient between Cr and Zr alloys and improve high temperature oxidation resistance better than Cr. The high temperature oxidation resistance of CrAl alloy was confirmed to be 10 times better than that of Cr, and CrAl-coated Zr alloy tube had excellent corrosion and creep resistance under normal conditions and maintained excellent deformation resistance and oxidation resistance even in LOCA simulation tests [12-13]. In the welded area of the coated pipe, no phenomena such as galvanic corrosion were observed under normal conditions, so the safety was confirmed to be good when welding the end cap of the coated pipe [12]. In addition, it was confirmed that there were no problems with soundness in a short-term study test in Halden research reactor [13].

Another advantage of a cladding tube coated with CrAl using AIP is that the amount of CRUD (Chalk River Unidentified Deposit) deposition is reduced compared to Zr cladding pipe. As a result of the simulation test as shown in Figure 6, it was confirmed that the amount of CRUD deposition during coating was small, and the CRUD layer was porous and easily separated [15]. These results show that the CrAl-coated cladding tube is expected to have the effect of alleviating CRUD deposition problems that occur in corrosive environments in commercial power plants, in addition to improving oxidation resistance.

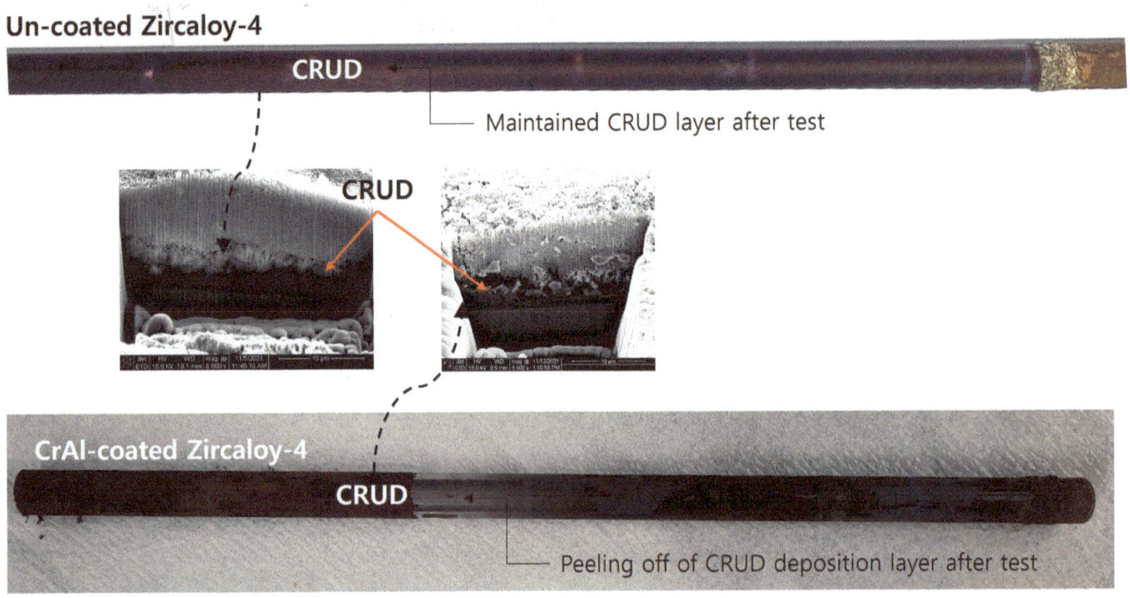

FIG. 6. Surface Appearance and Cross-sectional Observation of Un-coated Zircaloy-4 and CrAl-coated Zircaloy-4 Cladding by AIP Coating Method after CRUD Simulation Test.

However, while coating Cr or CrAl alloy using AIP technology, we made it possible to control the microstructure of the coating layer according to variables such as coating conditions such as chamber temperature, current, and voltage, as shown in Figure 7. However, it was found that the AIP coating variable causes a change in the microstructure of the internal zirconium alloy, which affects the bonding strength between the coating layer and the base material and the mechanical properties of the base material.

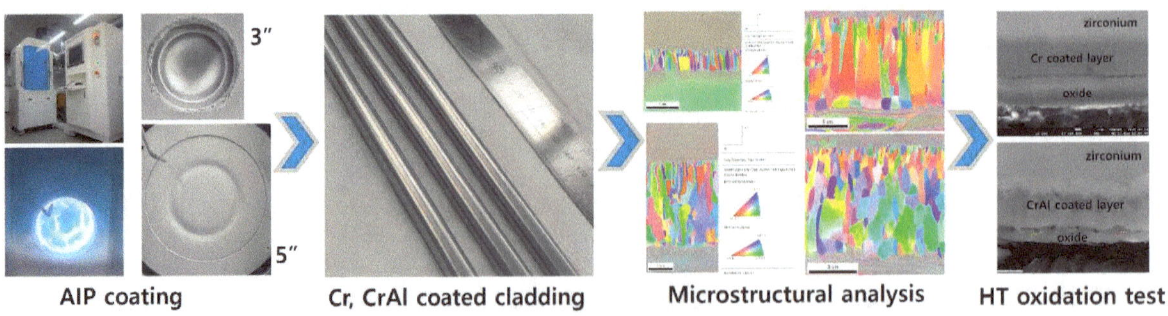

FIG. 7. Development Process of Cr and CrAl Alloy Coating Technology using AIP for ATF Zr Alloy Cladding, and High Temperature Oxidation Results of Cr and CrAl Alloy Coated Zr Cladding Development at KAERI

In this way, depending on the change in conditions during AIP coating, not only can the deposition rate of the coating film and the microstructure of the coating film be controlled, but also changes in the microstructure of the base material can occur. Increasing the deposition rate and improving the bonding strength are key requirements for AIP coating technology, but if the thermal energy due to the coating variable is increased for this purpose, the problem of recrystallization of the Zr alloy base material may occur, as shown in Figure 8. Recrystallization of these base materials creates a new problem of transforming the microstructure of Zr alloy cladding tubes (Zircaloy, ZIRLO^{TM}, HANA, etc.), which mostly have a stress relaxation heat treatment or partial recrystallization heat treatment structure, into a recrystallization structure. In other words, the recrystallization phenomenon lowers the fatigue strength by lowering the tensile strength, yield strength, and elastic modulus, and when there is a partial microstructure change, heterogeneous creep deformation, differences in irradiation growth, and differences in hydride direction occur, causing the cladding tube to be damaged. There is a risk that safety issues in normal operation conditions may arise. In addition, if the microstructure changes during the manufacture of an AIP-coated cladding tube, the problem of bending the cladding tube after coating may occur due to non-uniformity of stress. If a stress variation in the cladding tube is formed during AIP coating, the stress may cause a bowing problem during an irradiation test in a commercial reactor. In order to solve this problem, it is necessary to secure coating conditions with excellent adhesion without changes in the microstructure of Zr alloy substrate through evaluation and analysis of each variable applied during AIP coating.

KAERI has developed the optimized-AIP coating process that does not change the microstructure of Zr alloy cladding when coating Cr or CrAl alloy with AIP. Comparing the high temperature oxidation characteristics, the oxidation resistance of CrAl coating tube is very excellent compared to Cr, regarding the coating materials.

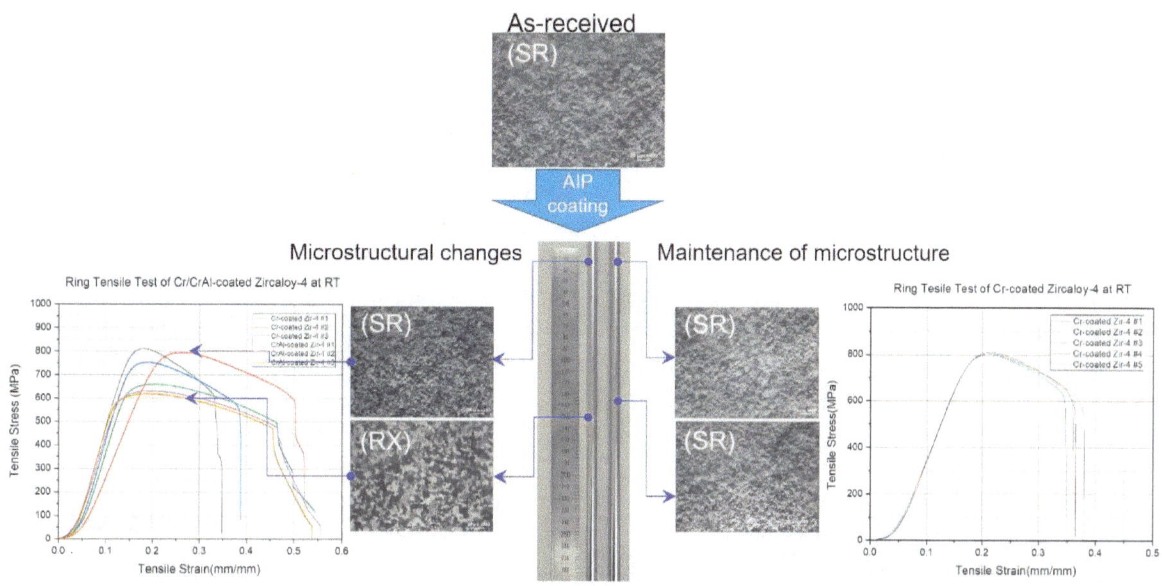

FIG. 8. Test Results Showing Changes in Microstructure and Strength According to Process Variables during AIP Coating.

3D printing technology has the advantage of being able to coat almost any material using the powder and has excellent bonding strength due to the formation of a diffusion layer between the coating material and the base material [7, 8]. However, because the cladding base material may have a problem in that its microstructure is partially changed by a high heat source, it is very important to constantly adjust the part where the microstructure changes (heat affected zone). Figure 9 shows the evaluation of coating an oxidation-resistant material using 3D printing technology. As a result of evaluating the properties by coating Cr, it was found that not only high-temperature oxidation resistance, but also high-temperature deformation resistance was greatly improved [7]. In addition, coating on zirconium cladding tube with FeCrAl provides excellent adhesion and increases high-temperature oxidation resistance, but it was confirmed that the inter-diffusion layer between FeCrAl material and Zr expands compared to Cr during high-temperature oxidation test [16].

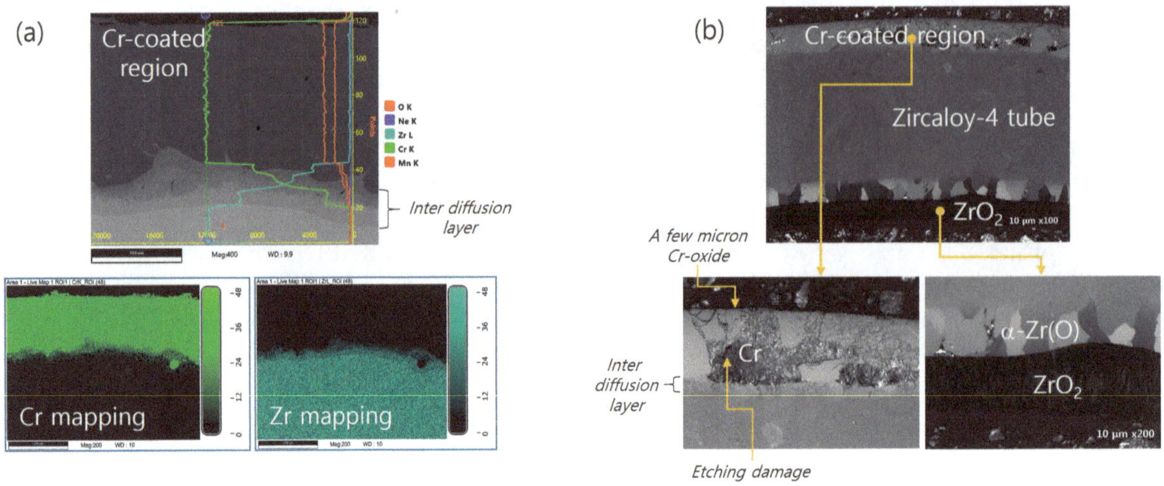

FIG. 9. Microstructural Characteristics and High-temperature Oxidation Behaviour of Cr-coated Cladding using 3D Laser Cladding Technology; (a) Cross-sectional Observation after Cr Coating and (b) Cross-sectional Observation after Oxidation Test of Cr-coated Cladding at 1200°C for 2000 s in a Steam.

Another advantage of 3D printing technology is that it is possible to improve the high-temperature strength of metal by strengthening oxide dispersion. When Y_2O_3 particles are dispersed to a thickness of about 80 μm on the surface of a zirconium cladding tube, the resistance to deformation and damage that occurs in the cladding tube during an accident such as LOCA is greatly increased, as shown in Figure 10. Due to this strength improvement effect, the increase in deformation resistance of zirconium cladding tube is expected to be advantageous in securing flow paths and have the effect of reducing FFRD (fuel fragmentation, relocation, and dispersal) problems caused by cladding deformation.

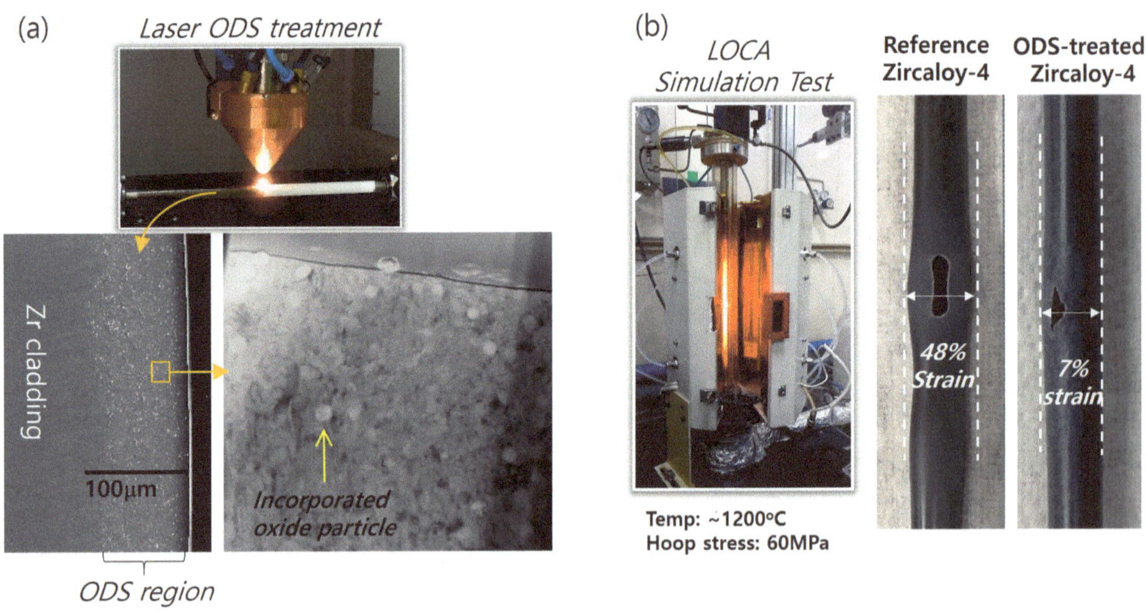

FIG. 10. Oxide Dispersion Strengthening (ODS) Treatment and High-temperature Properties on the Surface of Zircaloy-4 Cladding using 3D Laser Cladding Technology; (a) Laser ODS Treatment Method and Microstructure, and (b) Deformation Behaviour of ODS-treated Cladding during LOCA.

The cladding tube coated with FeCrAl using 3D printing technology was found to maintain durability without damage as a result of a short-term study in Halden research reactor, as shown in Figure 11. Since the tests conducted so far have a very short test time, there is a need to conduct long-term evaluation and post-irradiation testing of the evaluated specimens.

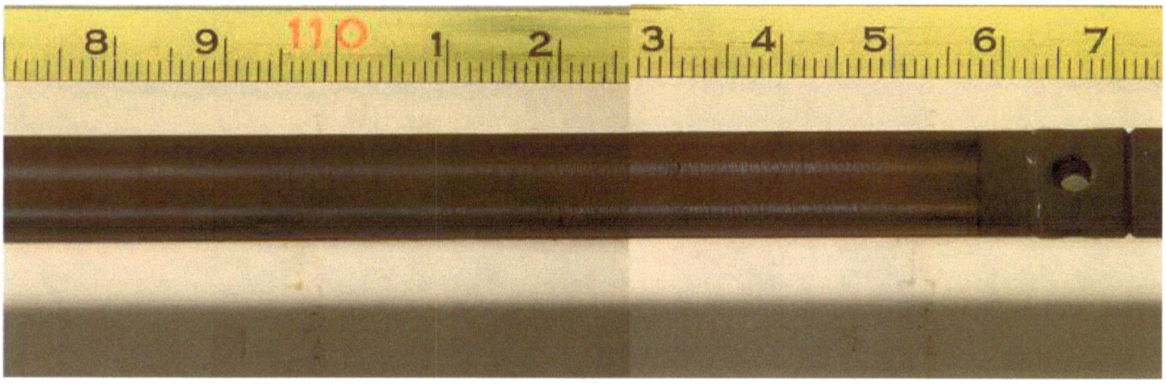

FIG. 11. Appearance of FeCrAl-coated Zircaloy-4 Cladding after One-cycle Irradiation Test in the Halden Reactor.

2.2.2 Summary of ATF cladding technology developed at KAERI for near-term applications

As materials for coating, Cr, CrAl, and FeCrAl, etc. were studied, and plasma spray, cold spray, AIP, and 3D printing technology were developed as coating methods. In order to improve high-temperature strength, an ODS structure was formed with 3D printing technology for Y_2O_3 particles. The performance of the surface-modified cladding was evaluated in comparison with the existing cladding through out-of pile and in-pile tests, and safety performance verification was also performed in case of an accident. The oxidation resistance of Zr alloy cladding in a high-temperature oxidizing atmosphere is improved by Cr, CrAl and FeCrAl alloy coatings, and the deformation resistance of Zr-alloy cladding in a high-temperature region is Improved by the formation of ODS structure.

All coating methods and materials applied to ATF cladding tubes developed as short-term technologies have advantages and disadvantages. When developing ATF cladding, researchers need to closely understand issues related to manufacturability, normal condition performance, accident condition performance, and disposal of spent nuclear fuel, and consider detailed items regarding applicability.

2.2 Development Status of 3D Printing Technology for Nuclear Applications

2.2.1 Considering 3D printing for nuclear applications

3D printing technology produces products by layering materials using three-dimensional modelling, contributing to expanding the design freedom of products by overcoming the limitations of existing manufacturing methods. In addition, in the case of discontinued parts, it has the advantage of being able to produce parts by inputting modelling data into a 3D printer without the need to restore lost manufacturing equipment. Due to these characteristics of 3D printing technology, it is being actively applied in various fields, and recently, its application is expanding in the nuclear field as well [8-11].

Recently, 3D printing technology is a manufacturing innovation technology that is being developed worldwide to be applied to various industrial fields, including nuclear energy. In particular, through exchanges with the nuclear industry, the US NRC reviewed 3D printing as necessary for nuclear reactor manufacturing due to its advantages in shortening the delivery period, improving design freedom for performance improvement, reducing production costs, and controlling chemical reactions in manufacturing, and together with related organizations, related regulations are being prepared [17-20].

KAERI's strategy for developing 3D printing (AM) technology is shown in Figure 12. The specific development strategy is to design the part shape for 3D printing (DfAM, Design for AM), manufacture prototypes, and verify quality through performance evaluation, securing core technologies that can be applied to various materials and processes. In addition, we are applying a technology development strategy that reflects the evaluation results and continuously improves the design of materials and manufacturing processes to improve the completeness of the results.

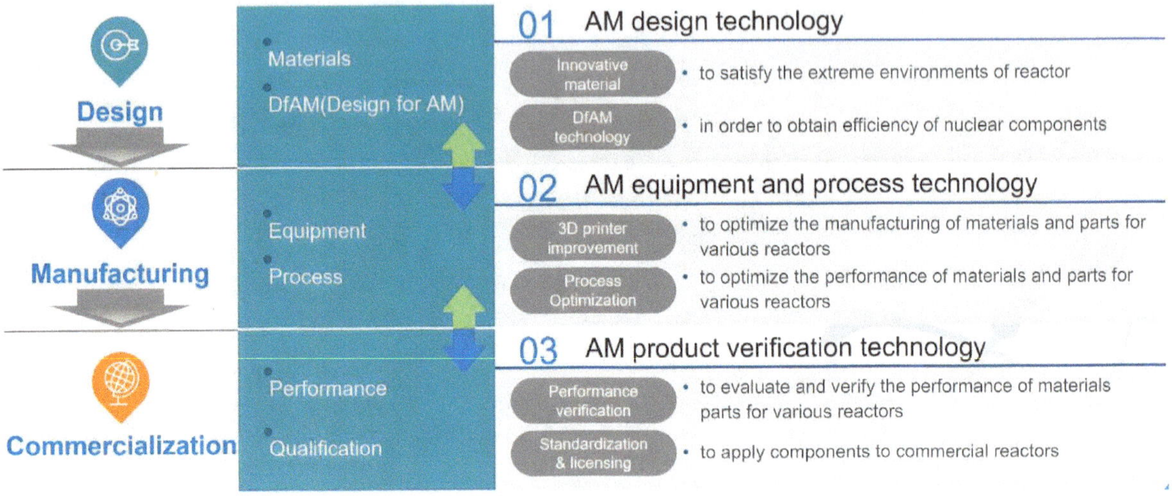

FIG. 12. Development Process of 3D Printing Technology at KAERI.

3D printing technology is an AM method, as shown in Figure 13, and includes technologies for simplifying the shape of parts and developing new shapes using known materials. In addition, it is possible to manufacture new materials that hybridize various materials by utilizing the characteristics of AM technology. Details for manufacturing components with innovative new material structures for application to nuclear power are as follows:

- Developing innovative materials and components using AM technology for nuclear reactors:
  - ✓ Complex structure manufacturing technology (PBF (Powder Bed Fusion)) to meet next reactor design;
  - ✓ Different material coating/cladding technology (DED (Direct Energy Deposition)) to increase reaction resistance and radiation shielding function;
  - ✓ Development of convergence technology of two AM methods (PBF+DED).

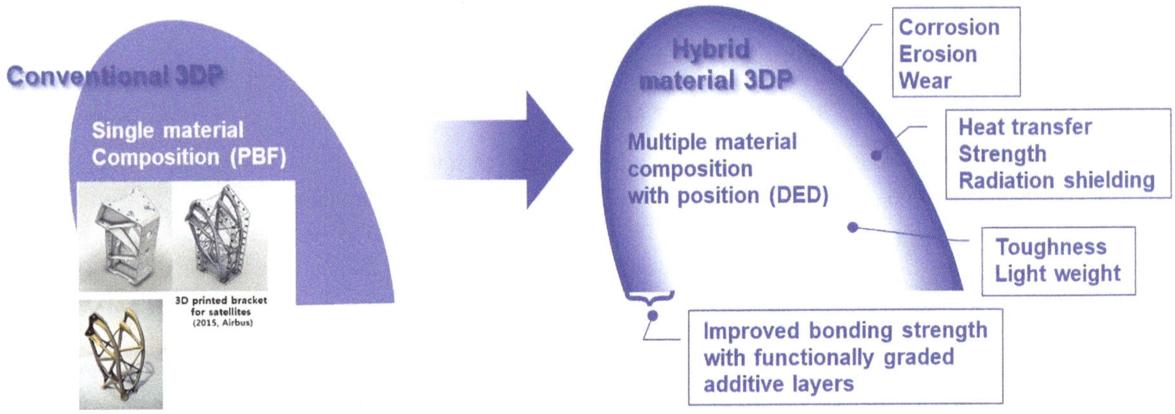

FIG. 13. Innovative Material and Component Development using 3Dprinting Technology at KAERI.

The technology developed by KAERI is PBF technology, which manufactures complex shapes without connections through 3D printing of nuclear materials, and DED technology, which increases the stability of parts exposed to specific environments by mixing and stacking various materials and combines the two technologies. We are also developing technology to restore damaged parts (ex. surface wear parts, etc.). It also secures technology for coating with the same material as the existing material and cladding with a material that can improve corrosion resistance and wear resistance compared to existing materials. This 3D printing technology can be used not only to restore damaged parts of operational NPPs and supply discontinued parts, but also to manufacture key material parts for small or ultra-small nuclear reactors.

2.2.2 Research results of the 3D printing technology developed at KAERI for nuclear applications

The 3D printing technology being developed by KAERI covers all aspects of new concept design, part moulding, and equipment and process technology for this.

Because 3D printing has a high degree of design freedom by stacking complex shapes, DfAM technology is applied for efficient design. As an example, KAERI is applying DfAM technology for topology optimization to improve the performance of heat exchangers and evaluate performance by manufacturing the designed drawings using the PBF method as shown in Figure 14 [22]. By applying this method, it is possible to apply a new heat exchanger concept that has better performance than the heat exchanger in the existing design. Heat exchangers designed with DfAM technology are expected to be highly applicable to small or ultra-small nuclear reactors because they are effective not only in improving performance but also in reducing the volume of components.

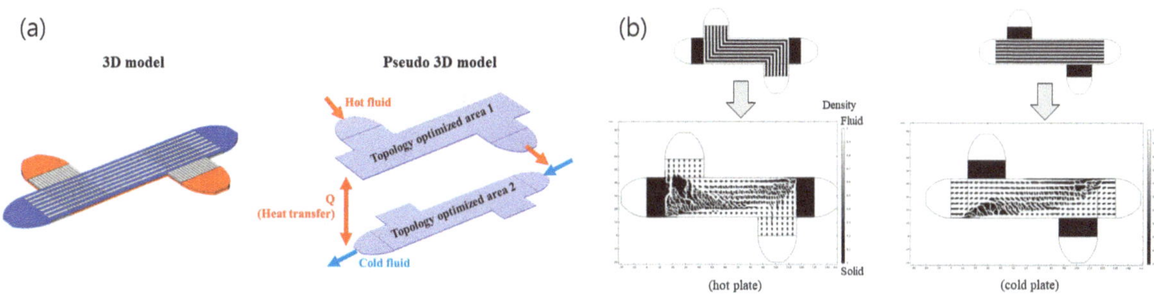

FIG. 14. Application of DfAM Technology for Optimal Design for Manufacturing Heat Exchanger by PBF Method 3D printing; (a) Pseudo 3D Method of the Optimization Algorithm and (b) Design Process for Optimal Phase Value.

Since nuclear materials and parts are composed of various materials, 3D printing equipment needs to be also flexible depending on the application target. KAERI developed PBF equipment to manufacture complex shapes from a single material, as shown in Figure 15, and designed and manufactured it to enable stacking of 1 m in length in the longitudinal direction. In order to expand the product size, a technology was developed and applied to connect a laser source, a scanner, and two powder supply devices in parallel. DED equipment for coating the surface of a part with the same or different materials or making parts for a specific purpose was manufactured by combining a computer numerical control based device capable of 5-axis movement with a powder supply device, nozzle, and laser. Another feature of this DED equipment is that it is possible to supply two or more powders during the process, which has the advantage of enabling layering of inclined functional materials or alloys of new compositions. As a method to quickly stack large parts, we also operate Wire-DED equipment, which is equipped with a head that can be fused with a laser while supplying wire to a multi-joint robot arm.

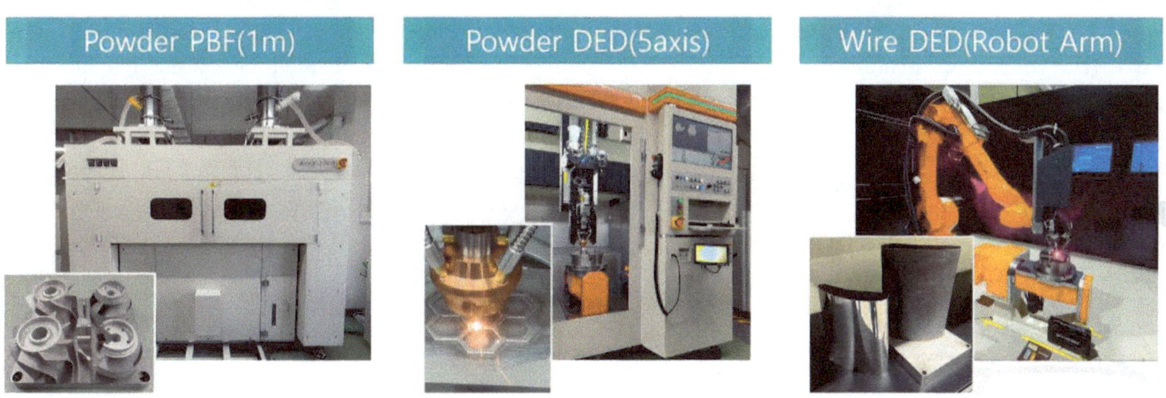

FIG. 15. Developing Various AM Devices after Considering the Material, Shape, Size, and Field Applicability of Parts/Components for Nuclear Application at KAERI.

Materials needed for the nuclear industry vary greatly depending on their purpose. Nuclear reactors are composed of ceramic nuclear fuel, high-temperature, high-corrosion-resistant core structural components, ultra-high temperature materials, radiation shielding materials, etc., and their form is also very complex depending on the design purpose of the nuclear reactor. To make various parts from these various materials, various manufacturing technologies are applied in combination depending on the purpose, size, shape, and manufacturing economy. Therefore, in order to secure nuclear power material parts manufacturing technology by applying 3D printing technology, AM and evaluation of nuclear power materials needs to be carried out.

KAERI has been developing equipment and process technology for manufacturing nuclear material components using 3D printing technology for the past 10 years. Because 3D printing manufactures parts through several steps such as melting, heat treatment, and processing of existing manufacturing technologies in one process, various AM variable requirements have to be controlled, as shown in Figure 16. Therefore, the core of AM technology is to derive process variables appropriate for each material and equipment, and the quality of manufactured items is to be maintained consistently. For this technology development, KAERI is developing manufacturing processes for each material that can be used in nuclear power by dividing them into DED method and PBF method. In addition, in technology development, we not only manufacture and evaluate products by process, but also identify heat and stress distribution through finite element method analysis and use them in process settings.

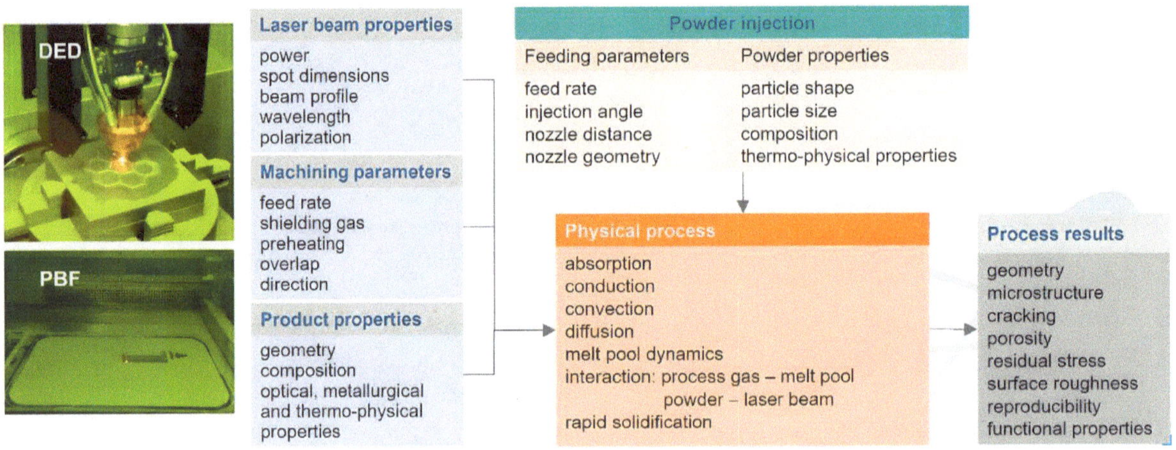

FIG. 16. Developing the Optimal AM Process for Each Material by Reviewing and Evaluating Important Variables for Nuclear Application at KAERI.

There are several examples of research and development conducted at KAERI to manufacture nuclear material components using AM as follows.

The study to optimize the stacking conditions of nuclear materials using the PBF method is shown in Figure 17. A method was followed to produce specimens applying various process variables, including laser output, and to derive optimal manufacturing conditions through physical property evaluation of the manufactured specimens. In this way, a process map for each material was implemented, and once the materials for nuclear power components were determined, the parts were manufactured under the optimal conditions identified in the process map. This method was equally applied to DED stacking. The process map for 3D printing will be applied to QC, and the quality of the manufactured product can be verified by producing, analysing, and evaluating evaluation specimens in the spare space during AM of parts.

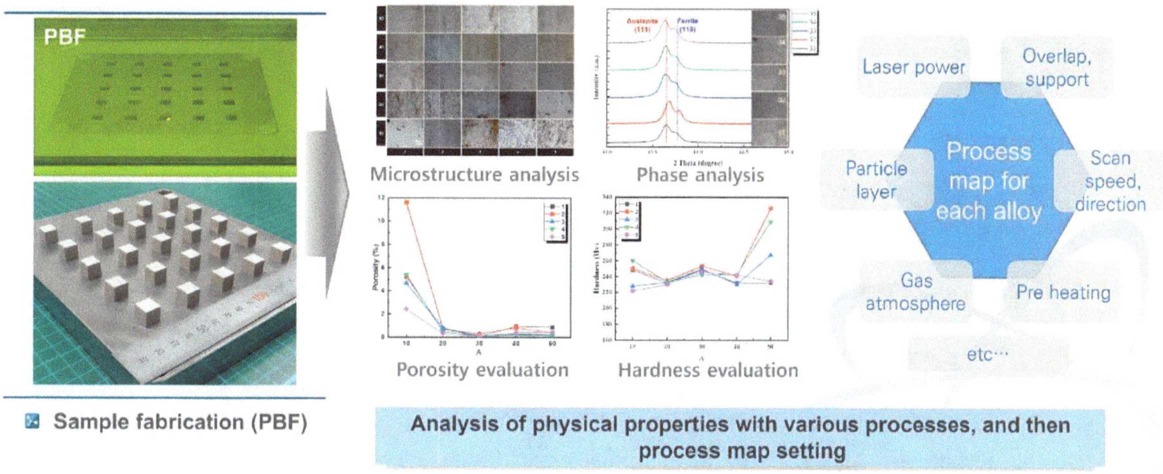

FIG. 17. Setting a Process Map for Nuclear Material to Obtain Optimal AM Conditions for PBF at KAERI.

Details related to AM and performance evaluation of complex shapes such as heat exchangers are shown in Figure 18. For the production of heat exchangers, DfAM-based structural design is performed to increase efficiency, 2-dimensional layers and tool paths are specified for 3-dimensional AM using CAD, and materials suitable for the heat exchanger are selected to determine the optimal process map as shown in Figure 16. A prototype was manufactured by applying the manufacturing process. Research and development were conducted by attaching the manufactured prototype to a heat exchanger evaluation device to verify its performance. The performance of heat exchangers manufactured through AM with the DfAM design was found to be improved compared to existing heat exchangers.

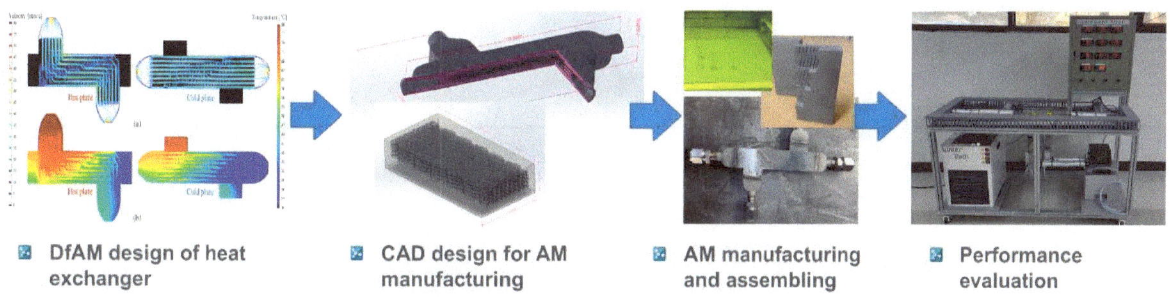

FIG. 18. Example of Development Process of Heat Exchanger using DfAM Design, CAD Design, Component Manufacturing (PBF), and Performance Evaluation for Nuclear Application at KAERI.

KAERI not only applies the DED method to restore a specific area of a damaged part to the same material and specifications, but also applies the technology of cladding different materials with the same concept as shown in Figure 13 to various materials as shown in Figure 19. The important point in the cladding of heterogeneous materials is to analyse the physical characteristics of each material and to establish the lamination conditions necessary for mutual reaction by analysing the mutual physical characteristics of the heterogeneous materials during cladding. If the cladding conditions are not suitable, problems such as delamination, cracks, and base metal components spreading to the cladding surface may occur, resulting in failure to meet the required performance. In the figure, SiC cladding on the metal base material was studied for the purpose of improving wear and corrosion resistance, B_4C cladding was studied for neutron shielding, W cladding was studied for radiation shielding and surface hardening, and Mo alloy cladding was studied for improving high-temperature strength. This cladding technology of different materials will be applied to the high-temperature safety, corrosion resistance, wear resistance, and shielding performance required for each component during the design and manufacturing of nuclear reactors.

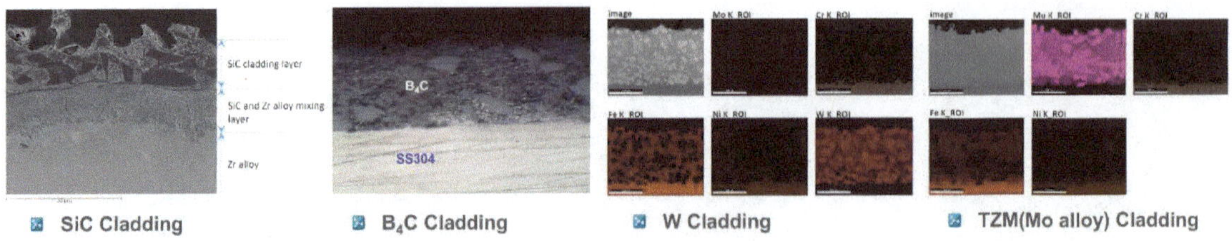

FIG. 19. Examples of Various Combinations of Cladding Technologies for Implementing Hybrid Materials Based on 3Dprinting Developed at KAERI.

2.2.3 *Summary of the 3D printing technology developed at KAERI for nuclear applications*

The 3D printing technology being developed by KAERI for application to the nuclear field includes design, part forming and analysis, and equipment and process technology for this. Through this, we are establishing PBF manufacturing technology and DED manufacturing technology for nuclear material components. In particular, we are developing technology that significantly improves performance compared to existing single-material components by hybridizing materials. If this innovative 3D printing technology is used to manufacture material components for new NPPs (SMR, micro modular reactor), it is expected to have various advantages such as improved design freedom, improved performance, restoration of damaged parts, and securing a supply chain for small quantities of various products.

In addition, 3D printed manufacturing material parts used in the extreme environment of nuclear power are not limited to nuclear power, as shown in Figure 20, but are highly applicable as spin-off technology to various industrial fields. However, there is still a need to establish an international system for ensuring reliability and quality certification for material parts manufactured with 3D printing technology.

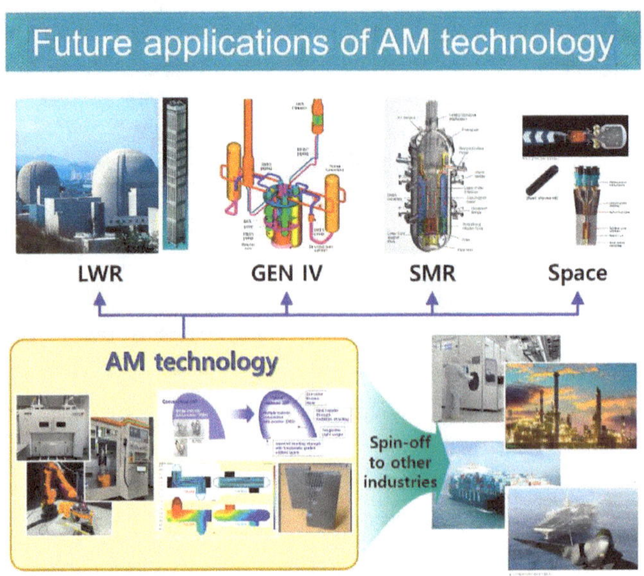

FIG. 20. Various Applications of 3Dprinting (AM) Technology in Nuclear as well as Other Industries for Extreme Environment.

3. CONCLUSIONS

KAERI has been developing surface modification technology for short-term ATF cladding to reduce the hydrogen explosion problem that occurred in the Fukushima accident, and 3D printing (AM) technology that can manufacture complex shaped parts and restore damaged parts.

Regarding the ATF, KAERI is systematically building technologies related to improving safety and economic efficiency and commercializing ATF cladding tubes incorporating coating methods as a short term technology. In particular, Cr or CrAl alloy was coated using the AIP process without damaging the base material, and evaluated in normal and accident environments, showing that it has superior performance compared to the current zirconium alloy in the event of an accident.

In addition, KAERI is strategically developing the core technologies needed to manufacture and apply new materials and components for nuclear power by applying 3D printing technology, which has recently become an issue. The innovative 3D printing technology developed by KAERI can be used to improve design freedom, improve performance, restore damaged parts, and secure the supply chain for small quantities or multiple types of products when applied to the manufacturing of material components for new NPPs, such as LWR, SMR and Micro modular reactors.

ACKNOWLEDGEMENT

This work was supported by a National Research Foundation of Korea (NRF) grant funded by the Korea Government (RS-2022-00144289), and was supported by the KAERI R&D Programs [grant numbers KAERI-524490-23].

REFERENCES

[1] ZINKLE S., et al., Accident Tolerant Fuels for LWRs: A Perspective, J. Nucl. Mater. **448** (2014) 374-379.

[2] KIM., H. G., et al., Application of Coating Technology on Zirconium-Based Alloy to Decrease High-Temperature Oxidation, in Zirconium in the Nuclear Industry: 17th International Symposium, ed. B. Comstock and P. Barberis (West Conshohocken, PA: ASTM International (2014) 346-369.

[3] PARK., J.H., et al., High temperature steam-oxidation behavior of arc ion plated Cr coatings for accident tolerant fuel claddings, Surf. Coat. Technology **280** (2015) 256-259.

[4] BISCHOFF., J., et al., AREVA NP's Enhanced Accident-Tolerant Fuel Developments: Focus in Cr-Coated M5 Cladding, Nucl. Eng. Technol. **50** (2018) 223-228.

[5] OELRICH., R., et al., Overview of Westinghouse Lead Accident Tolerant Fuel Program, Top Fuel 2018, Prague (2018).

[6] NUCLEAR ENERGY AGENCY, State-of-the-Art Report on Light Water Reactor Accident-Tolerant Fuels, OECD 2018, NEA No. 7317, Paris (2018).

[7] KIM H. G., et al., Adhesion property and high-temperature oxidation behavior of Cr-coated Zircaloy-4 cladding tube prepared by 3D laser coating, Journal of Nuclear Materials **465** (2015) 531-539.

[8] KIM H. G., et al., Microstructure and mechanical characteristics of surface oxide dispersion-strengthened Zircaloy-4 cladding tube, Additive Manufacturing **22** (2018) 75-88.

[9] WORLD NUCLEAR NEWS, 3D-printed fuel parts complete initial irradiation cycle (Framatome), World Nuclear News (2020).

[10] WORLD NUCLEAR NEWS, Westinghouse unveils 'industry first' with 3D-printed component, World Nuclear News (2020).

[11] USNC licenses 3D printing for reactor component manufacture, World Nuclear News 11 January 2022.

[12] KIM., H. G., et al., Out-of-pile performance of surface-modified Zr cladding for accident tolerant fuel in LWRs, J. Nucl. Mater. **510** (2018) 93-99.

[13] KIM., H. G., et al., Development of Cr-Al Coating on Zircaloy-4 for Enhanced Accident Tolerant Fuel, in Zirconium in the Nuclear Industry: 19th International Symposium, ed. A. T. Motta and S. K. Yagnik (West Conshohocken, PA: ASTM International (2021) 172-188.

[14] PARK., D.J., et al., Microstructure and mechanical behavior of Zr substrates coated with FeCrAl and Mo by cold-spraying, J. of Nucl. Mater. **504** (2018) 261-266.

[15] KIM., S. E., Enhanced crud resistance of CrAl coated ATF claddings in simulated PWR conditions, J. Nucl. Mater. **578** (2023) 154357.

[16] KIM., I.H., et al., Oxidation-resistant coating of FeCrAl on Zr-alloy tubes using 3D printing direct energy deposition, Surface & Coatings Technology **411** (2021) 126915.

[17] NUREG/CP-0310, Proceedings of the Public Meeting on Additive Manufacturing for Reactor Materials and Components.

[18] US NRC, Draft Guidelines Document for Additive Manufacturing – LPBF, ML21074A

[19] ALBERT., M., Vision of Advanced Manufacturing Technology (AMT) Use in the Nuclear Industry, NRC Workshop on Advanced Manufacturing Technologies for Nuclear Applications (2020).

[20] PUUKKO., P., "QA and QC Tools for Metal AM and implementing them in EU NUCOBAM project", NRC Workshop on Advanced Manufacturing Technologies for Nuclear Applications (2020).

[21] KIM., S. E., et al., An Optimization of Manufacturing Process of SS304L Using Direct Energy Deposition Method, Korean J. Met. Mater., **60** 11 (2022) 819-826.

[22] LEE, G. H., et al., Dual-fluid topology optimization of printed-circuit heat exchanger with low-pumping-power design, Case Studies in Thermal Engineering **49** (2013) 103318.

TECHNICAL SESSION IV: ADVANCEMENTS IN FABRICATION FACILITES AND INSPECTION

ID#15

DEVELOPMENT TRENDS AND RECENT ACHIEVEMENTS IN FUEL FABRICATION TECHNOLOGY AT TVEL FACILITIES

A. RADOSTIN
TVEL,
Moscow, Russian Federation

Abstract

The paper describes the development trends and recent achievements in fuel fabrication technology at Rosatom fuel company (TVEL) facilities. The TVEL fuel feeds more than 70 power reactors and 30 research reactors in thirteen countries. TVEL possesses the leading competence in the design and manufacturing of nuclear fuel for the LWR and continues developing them.

1. INTRODUCTION

Fuel Company TVEL is designing and producing nuclear fuel and fuel components for power and research reactors. Traditionally TVEL has been producing nuclear fuel for the reactors developed in the Soviet Union and Russia. Nowadays TVEL manufactures fuel and fuel components for power and research reactors of Western design as well. The TVEL fuel feeds more than 70 power reactors and 30 research reactors in thirteen countries. Thus, TVEL is among the leading global suppliers of nuclear fuel.

TVEL combines enterprises for manufacturing gas centrifuges, uranium conversion and enrichment, nuclear fuel fabrication, as well as research and development organizations. TVEL comprises and manages all the nuclear fuel cycle capacities in Russia except for uranium mining. TVEL's products are being supplied by TVEL directly except for Enriched Uranium Product and separative work units which are supplied via Tenex.

Thus, TVEL is a single multidisciplinary scientific and production system. The chain of engineering, supply, and service of the nuclear fuel created by the long-term work of TVEL also includes a number of the leading Russian and foreign companies and relies on Rosatom's international business infrastructure, which ensures secure, uninterrupted support for the operation of nuclear reactors — users of the fuel manufactured by TVEL. The place of TVEL in the Nuclear Fuel Cycle is shown in Figure 1.

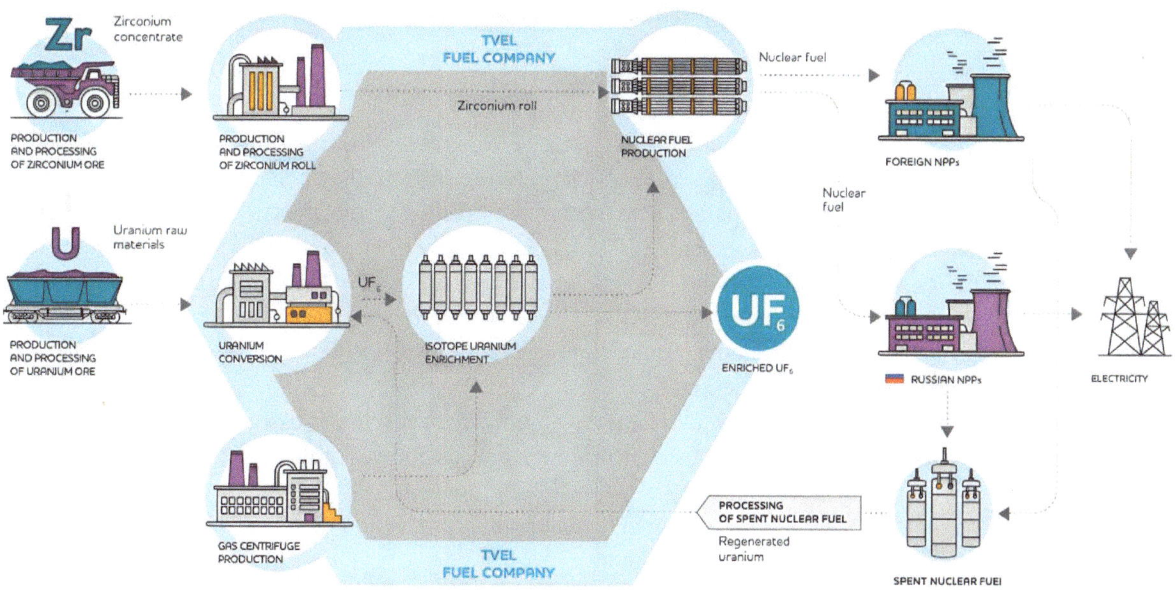

FIG. 1. TVEL in Nuclear Fuel Cycle.

Security of supply is provided by reservation of capacities. TVEL has four enrichment facilities (ECP – Zelenogorsk, AECC – Angarsk, SCC – Seversk and UEIP – Novouralsk) and two fuel fabrication facilities (NCCP – Novosibirsk and MSZ – Elektrostal). TVEL also practices technology transfer where feasible. For example, WWER fuel is also fabricated to our design in China for the local NPPs.

Continuous development and improvement of nuclear fuel designs and processes are achieved by building long-term relationships with several leading design companies (e.g. 'OKBM Afrikantov', JSC and 'OKB GIDROPRESS', JSC) in Russia that have a high international profile, including, where appropriate, elements of competition.

This reasonable balance of integration and diversification of value chains provides a broader and more comprehensive asset management, covering a whole range of production areas and engineering solutions. This allows for the concentration of resources and ensures a broad exchange of best practices. A result is the enhancement of the corporate knowledge base.

TVEL possesses the leading competence in the design and manufacturing of robust nuclear fuel for LWRs. The materials and design solutions involved build up 30% stiffer WWER and PWR fuel than competing offers providing high efficiency and corrosion resistance of materials [1]. The pellet design has been based on larger grain size within the standard UO_2 specification reducing the fission gas release during operation at high burnup. These characteristics have been proven by decades of successful in-core performance in reactors of various designs.

Nuclear fuel at TVEL is working in the system of regulations ranging from intergovernmental agreements and basic national legislation to contracts and process documents. The system covers the whole life cycle of production facilities.

TVEL operates under ISO and NQA-1 quality systems. In general, the integrated management system allows for flexible customization of the product and supporting processes to meet the needs of a particular customer.

The technical policy towards the life cycle of fabrication equipment is based on the following principles:

- Comprehensive management of condition and advancement of the technical systems;
- Minimization of total cost of ownership;
- Continuous advancement of processes;
- Refreshment and modernization of fixed assets in accordance with the Russian and global best practices;
- Unification of processes and technical solutions, unification of the new equipment with the operable;
- Application of modern methods of maintenance and repair of equipment;
- Absolute priority of stable quality and safety of manufacturing.

2. MODERNIZATION OF THE FUEL FABRICATION FACILITIES AND ADAPTATION TO PWR FUEL

There are two fabrication facilities within TVEL: MSZ (Elektrostal) – the plant was founded in 1917, and NCCP (Novosibirsk) – the plant was founded in 1948. During their history, they underwent several significant rearrangements and upgrades. The latest upgrade of the fabrication facility in Electrostal was performed in 2019. The plant in Novosibirsk finalized refurbishment in 2021 including the commissioning of a series production line of TVSK fuel for PWRs. TVSK is the Russian 17 × 17 square profile FAs for PWRs. The completed projects resulted in the following effects: the production areas were reduced, several operations were optimized, which led to significant savings in stocks, and the automation of some operations was increased, which led to a decrease in the influence of the human factor on production.

Automation of pellet inspection (Figure 2) is aimed at eliminating inspection by operators using glove boxes. Inspection is carried out using video cameras that perform high-speed shooting of the pellet at the circumference and end faces.

A result was a significant performance improvement in this production area and the elimination of the inspection quality dependence on the work of operators.

Modernization of the pellet- loading-in-container area was aimed at reducing the risk of pellet damage during movement to the fuel rod production area and at reducing the amount of work in progress.

The renewal of the fuel rod production lines was aimed at reducing the number of movements of the components and finished products.

As a result, compact high-performance automated fuel rod production lines have been built at both facilities. Fuel rods for WWER-440, WWER-1000, and WWER-1200 reactors, as well as PWR 12 ft and 14 ft can be manufactured at these lines.

FIG. 2. Automated pellet inspection.

Automated fuel rod production lines include state-of-the-art production and inspection systems that ensure fuel rod fabrication from feeding the fuel rod cladding to the line to packing the finished fuel rods into a transport container.

QC carried out at each stage of fabrication enables making sure of the required quality of a fuel rod, and its compliance with the specification requirements.

Assembly components that are part of the FA are produced at special benches.

FAs are assembled at the assembly lines, where, along with the special benches, automated systems are used to ensure the accurate mutual arrangement of the components during assembly. When organizing workplaces for the manufacture of skeletons (Figure 3), assembly, inspection, and packaging of the FAs, the principle of a single process flow (sequence) with the pellet loading line is implemented.

FIG. 3 Stand of skeleton welding.

The described upgrade programmes were aimed at feeding the rapidly growing global WWER fleet and increasing demand in the PWR market. The next stage is the introduction of ATF solutions, production of higher efficiency fuel (providing for longer cycles, up to 24 months), and establishment of remote fabrication for U–Pu fuel.

Management of the equipment life cycle is based on the following principles:

- Use of reliable and high-performance equipment from domestic and international original equipment manufacturers;
- Exchange experience and best practices with peers in the global nuclear industry and other industries;
- Introduction of Total Productive Maintenance methods for maintenance and repair of equipment;
- Increase of equipment dependability and period between overhauls: selection of efficient materials and parts; advancement of processes; monitoring of equipment condition in operation, performance, time in service; predictive maintenance.

3. CONCLUSION

The key technology changes for WWER and PWR fuel fabrication are listed below:

- Automated pellet take-off from press;
- Pellets loading on pallets;
- Automated pellet inspection;
- Fuel rod loading with pellets from pallets;
- Laser marking of claddings;
- Automation of rod bundle mounting;
- Automation of FA skeleton welding;
- Establishment of U-Gd fuel fabrication area;
- Establishment of a single flow of fuel rod and FA mounting.

REFERENCES

[1] RADOSTIN., A., et al., Confirmation of the design characteristics of the TVS-K design after operation in the PWR reactor at Ringhals-3 NPP, the 2023 Water Reactor Fuel Performance Meeting (WRFPM2023), Xi'an, China, July 17-21, 2023.

ID#16
IMPROVEMENT OF HOMEMADE COATER

D. FAN, H. ZHANG, W. MI, X. LIANG, L. XI
Fuel Rod Workshop,
China North Nuclear Fuel Corp., Ltd,
Baotou, China

Abstract

In the process of equipment experiment and normal production, operational experience was accumulated by exploring the performance of equipment components. At the same time, uncertainties affecting process indicators and equipment stability are collected and recorded to improve and optimize step by step. Improve the stability, safety, and operability of the equipment.

Key words: Nuclear Fuel, Coater

1. INTRODUCTION

In order to ensure the digestion and absorption of the imported technology, develop the capacity to produce AP1000 fuel elements, and prepare for the large scale application of the AP1000 fuel type in China, CNNFC has successfully localized the AP1000 fuel pellet coating equipment through collaborative efforts of all parties involved.

In this study, the homemade multi-target direct current magnetron sputtering coater (hereinafter referred to as the 'homemade coater'). Using UO_2 pellets as the matrix and ZrB_2 as the sputtering target material [1], research was conducted on domestic coating furnace equipment. Based on the operator feedback and statistical analysis of abnormal operating conditions of equipment, the homemade coater was improved and optimized to improve the coating quality stability and improve the operation reliability of the homemade coater.

2. PROCESS PRINCIPLE

Magnetron sputtering [2] is a complex process involving the transfer of energy and momentum during collisions between ions and atoms on the material surface, resulting in the excitation of target atoms (Figure 1).

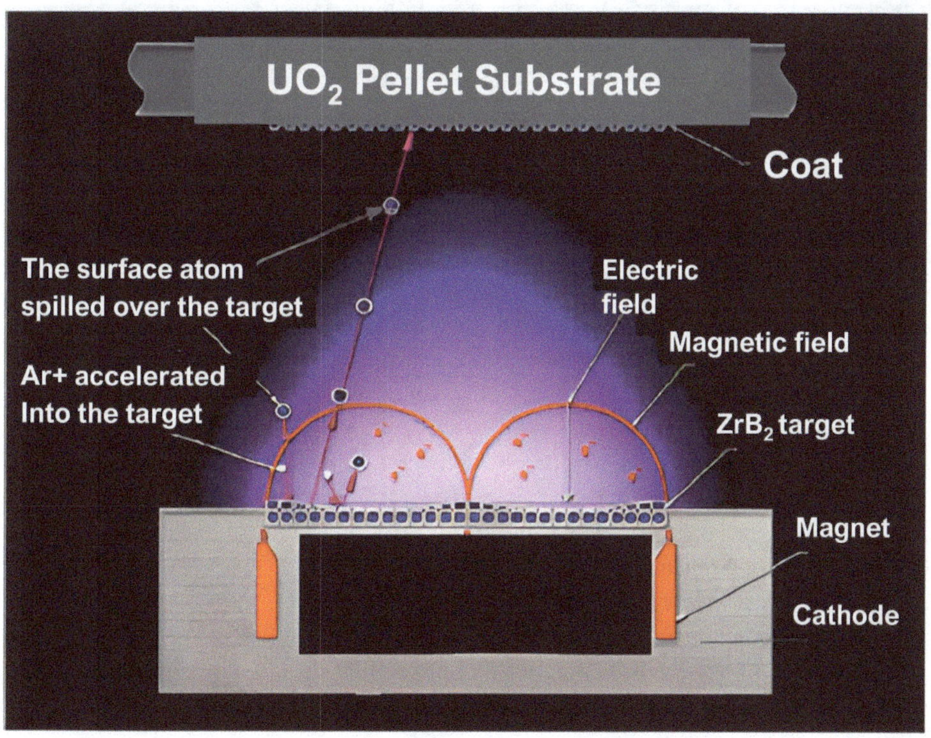

FIG. 1. Magnetron sputtering schematic diagram.

3. EQUIPMENT IMPROVEMENT

3.1 Equipment

The homemade multi-target direct current magnetron sputtering coater comprises a main unit (Figure 2 (a)), a vacuum system, a cathode power system, a cold and hot water-cycling system for target cooling, and a control system, among other components. (Figure 2 (b) and Figure 3).

(a) (b)

FIG. 2. Main part(a) and Auxiliary system (b).

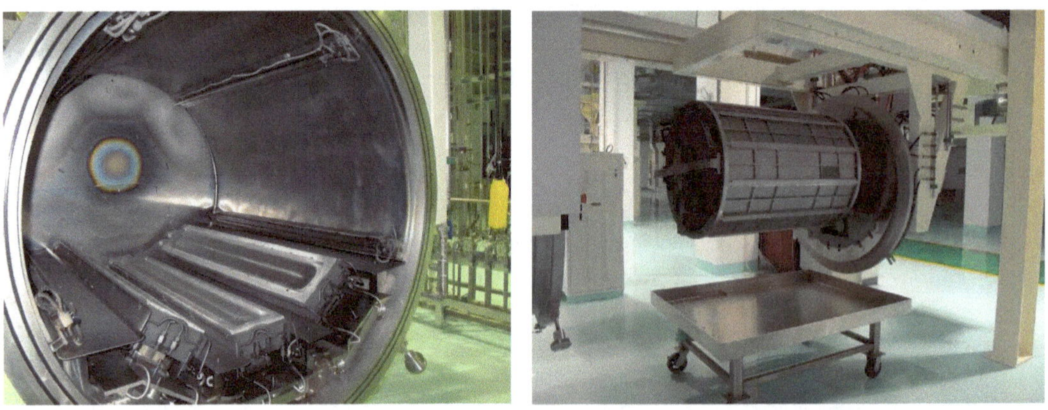

FIG. 3. Circular drum and six cathodes.

3.2 Equipment status

The main issues encountered with the use of the homemade coater are as follows:

— Drum Sliding and associated risks:
  Under the influence of the material and drum, the drum tends to slide toward the inside of the furnace. In severe cases, this may cause the drum to fall off, leading to damage to equipment components, product defects, and potentially causing collisions with personnel or further equipment damage.

— Bellow Design Issues:
  The bellows occupy a large amount of space, with thin tube walls and close proximity to the edge of the cathode. During the initiation process, this design can lead to a short circuit and the punch-through incidents, resulting in water leakage or significantly reduced coating efficiency.

— PLC Control System Problems:
  - Cathode power interface display errors;
  - No early-warning function for frozen water in helium compressor;
  - The automatic program fails to progress to the specified stage according to the design requirements;
  - The target material has a short lifespan, and the total kWh is not accumulated;
  - Data is cleared when restarting the program after a cathode disconnection.

3.3 Improvement plan

3.3.1 *Improvement of drum spindle*

The structure of the drum core shaft has been improved. The design is shown in Figure 4:

— The drum spindle structure was modified by adding a limit platform to the original groove design. This ensures that the drum will not idle when it stops.

— A limit block was added at the connection point between the drum spindle and the drum motor to prevent the drum from sliding inward:

- The outsourced machining spindle underwent an applicability test to evaluate the operating state of the equipment;
- Improvements were made based on the test results until the design requirements were met.

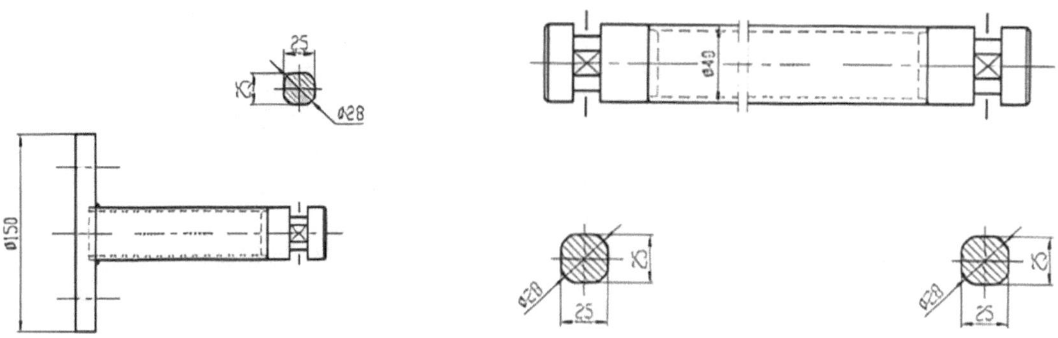

FIG.4. Drum spindle design drawing.

3.3.2 *Improvement of cathodic shielding plate water pipeline*

1) An L-shaped water pipe structure was developed to ensure the cooling function for the cathode while reducing the pipe height to eliminate the risk of pipeline breakdown; The pipe joint was positioned parallel to the wall welding spot;

2) After these improvements, a suitability test was conducted on the sample;

3) Further improvements were made based on the test results until the design requirements were met.

3.3.3 *System improvement*

Several automation issues in the PC were identified and resolved through hardware and software modifications:

— Cathode Power Interface Display Error:
The hardware connection and PLC software construction were checked and optimized.

— Lack of Early-Warning Function for Frozen Water in the Helium Compressor:
Differential pressure gauges were installed at the inlet and outlet of the filter unit for hardware improvements. In software, differential pressure restriction conditions and an alarm interlock function were added. The automatic sputtering procedure now halts when pressure remains constant, ensuring product quality and equipment operation safety.

— Automatic Program Failing to Progress to Specified Stages:
Online monitoring PLC program operation was introduced to identify problematic segments and implement corrective measures.

— Short Life of Target Material and Lack of Total kWh Accumulation:
An alarm function was added to the PLC system to notify when the target material's lifespan is nearing its end, interlocking with the automatic program. When an alarm is triggered, the PLC program restricts the start condition of the cathode power.

— Data Clearing After Program Restart Following Cathode Disconnection:
The PLC data reading and writing rules were optimized to prevent data loss during program restarts.

4. IMPROVEMENT RESULTS AND ANALYSIS

4.1 Drum spindle

Considering the stability of the structure and the collocation degree of the components, the original structure was modified to reduce the costs and improve the safety of the equipment. After multiple revisions to the design drawings, the equipment was assembled on site. The rework of the original part and addition of a limit platform to the original slot design were successfully completed (Figures 5 and 6).

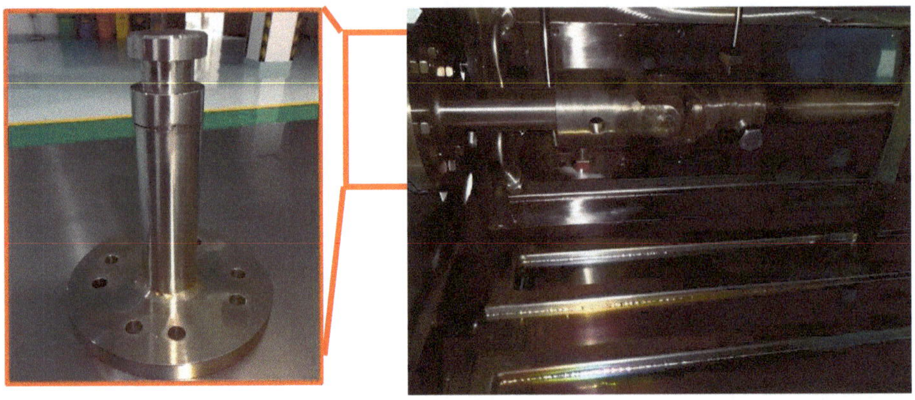

FIG. 5. Drum spindle.

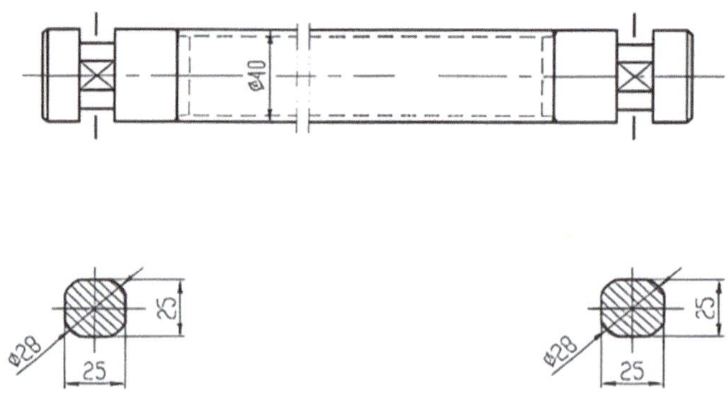

FIG. 6. Improvement drawing of drum spindle.

4.2 Cathodic shielding plate water pipeline

The improvement was carried out in two ways:

1) A small number of original pipes were purchased to maintain a reserve, minimizing costs;

2) Customized spare parts of the same model were created to ensure consistency in product quality and compatibility with existing materials.

In the second method, an L-shaped water pipe structure was adopted. The design ensures the cooling function of water pipe while reducing its height to eliminate the risk of sputtering breakdown (Figure 7). To mitigate shear forces acting on solder joints during disassembly and installation, the pipe joints were placed parallel to the wall solder joints, reducing the likelihood of damage to the solder joints.

After completing the pipeline customization, load testing on the equipment confirmed that the new L-shaped pipeline did not negatively affect the equipment or products. Thus, the cathodic shielding plate water pipeline improvement was completed. The frequency of problems in recent years is shown in Figure 8.

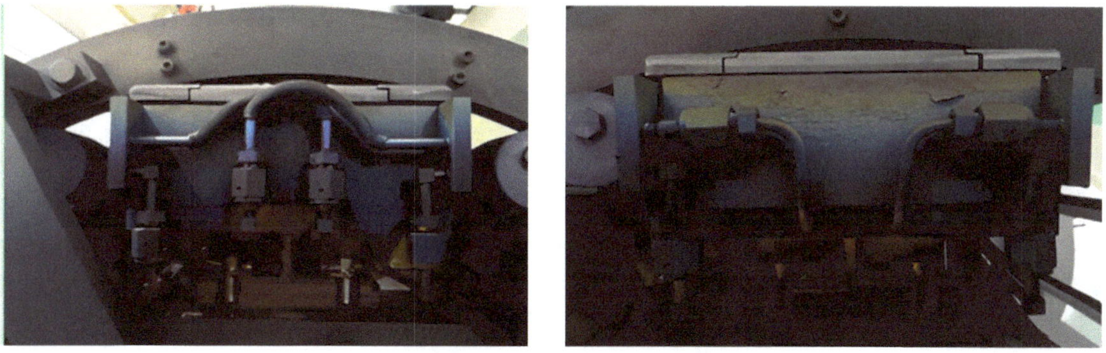

FIG. 7. Before and after improvement of pipe.

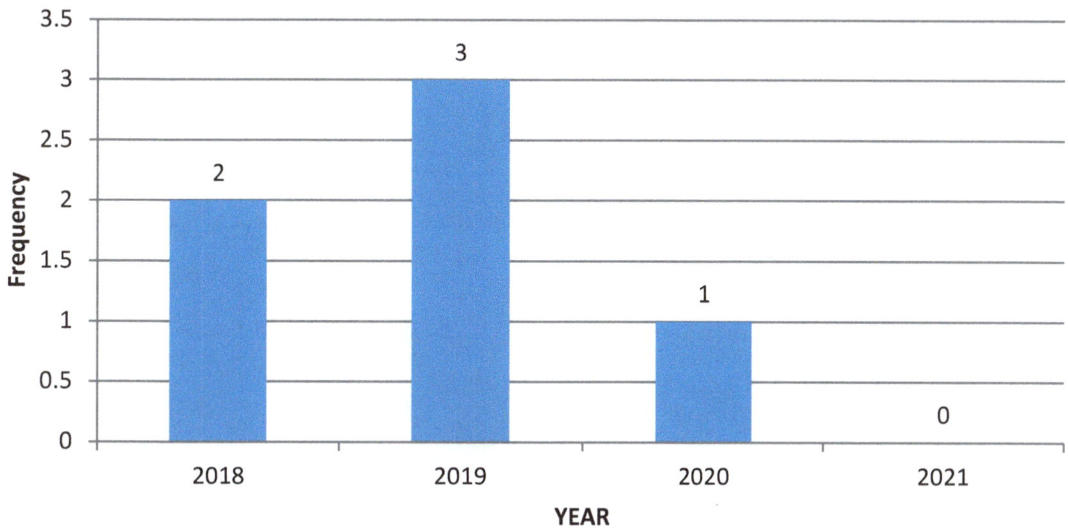

FIG .8. The frequency of problems.

4.3 System improvement

1) Cathode Power Interface Display Error:

Analysis revealed that sputtering power is provided by a DC high-voltage cathode. Occasionally, short millisecond on-off shocks occur due to interrupted output power, leading to abnormal kWh accumulation. The accumulated data becomes excessively large, exceeding display limits and causing errors like "####."

To address this issue, software was added to suppress on-off shocks of the output flag bit. When the power is off, lock M7.1 is immediately switched on, ensuring that kWh accumulation is unaffected.

2) No Warning Function for Frozen Water in the Helium Compressor:

Differential pressure gauges were installed at the inlet and outlet of the filter device. Differential pressure restriction conditions and an alarm interlock function were implemented in the software. When the pressure difference remains constant, the automatic sputtering procedure is halted to ensure product quality and equipment safety.

3) Automatic Procedure Failing to Progress to the Specified Phase:

Investigation showed that the program stops at the end of the pressure stage due to a large temperature difference in the cold and hot water switch design. Slow water temperature increases prevented the program from reaching the set temperature within the designated time, reducing production efficiency.

To resolve this, the temperature settings were modified within regulatory limits. A user-friendly modification interface was added to allow future adjustments.

4) Short Target Material Lifespan Due to Unmeasured Total kWh:

When the target material lifespan drops below the set total kWh, the cathode power supply remains active during sputtering, but the kWh is not recorded. This can lead to excess boron coating and product defects if operators fail to monitor accumulated energy on the cathode.

The PLC program was updated to address this issue. When the target life falls below 100 kWh, an alarm is triggered, and the coating program halts to maintain product quality.

5) Data Clearing After Program Restart Following Cathode Disconnection:

Two solutions were proposed:

- A judgment program compares the kWh across cathodes at startup. If the difference between one cathode and others exceeds a set limit, an alarm is triggered, and the program halts;
- A zero clearing function was added for all cathode KWH values. Operators manually record revised kWh, and a Visual Basic for Applications (VBA) program calculates the cumulative remaining kWh.

The second solution was adopted. All cathode values are cleared to zero, ensuring accurate cumulative kWh calculation.

Statistics on system problem frequency in recent years are shown in Figure 9, demonstrating the success of these improvements.

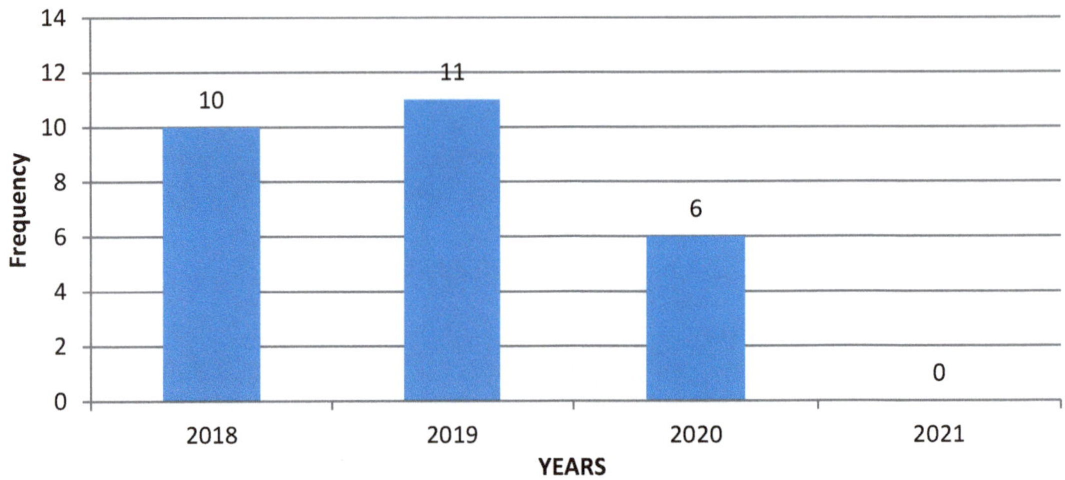

FIG. 9. The frequency of system problems.

5. CONCLUSIONS

1) Controlling instability in the mechanical components of the homemade coater and addressing factors affecting operation safety have significantly improved equipment safety. Through investigation, trial production, verification and application, cathode improvements and localization have been achieved, enhancing the spare parts supply chain's stability and reducing dependence on imported components;

2) Optimizing the hardware and software of the homemade coater has improved its operational stability and usability, providing a reliable foundation for consistently producing high-quality products.

REFERENCES

[1] LUI, R., Brief Introduction to Magnetron Sputtering Coating Technology. Science and Technology Frontier. (2006) 56-59 (in Chinese).

[2] XU, W., Progress and application of magnetron sputtering technology. Beijing. Modern Instruments **5** (2005) 1-3 (in Chinese).

ID#18

NEW TECHNOLOGIES AND INNOVATIONS IN QUALITY INSPECTION EQUIPMENT AT ENUSA'S FACTORY

D. VERDEJO GARRIDO, J. CASTAÑO MARCOS
New Technologies and Equipment Development Engineers,
ENUSA Industrias Avanzadas S.A. S.M.E. (Grupo SEPI),
Juzbado, Spain

Abstract

This paper provides a general overview of the new technologies that ENUSA is implementing in the fuel manufacturing process to improve the process itself and its results.

Nuclear fuel manufacturing, like other processes in the nuclear industry, is characterized by two main aspects: safety and quality. This requires a compromise between already known and reliable technologies, and high performance and cutting-edge ones.

The main objective is to manufacture nuclear fuel with high-quality standards, in strict compliance with the safety rules, and trying to do it as time and cost effective as possible. To achieve this, ENUSA is continuously improving its processes and equipment, especially the most advanced and complex ones.

This paper covers the innovation technologies that are being applied to two of the most advanced inspection equipment in ENUSA's factory: the use of Deep Neural Networks in API; and rod surface inspection by laser profiler (3DM-Pro). Also, the application of Augmented Reality on some other processes around the manufacturing process (control, monitoring, maintenance, etc.).

1. INTRODUCTION

Nowadays, all industries are adopting new and advanced technologies in all types of processes. The emergence and evolution of these new technologies are ever more rapid, so their implementation in industries need to be continuous and fast.

Nuclear industry is characterized by two main aspects: safety and quality. Safety, from the point of view of technology usage, causes relatively low implementation of cutting-edge technologies. The technologies used in all the steps of nuclear cycle, ranging from conversion process to spent nuclear fuel management and storage, including fuel fabrication, power generation on the reactors, reprocessing, etc., needs to be already proven and reliable.

On the other hand, the high-quality standards required in nuclear industry, push it to adopt high performance technologies for monitoring, measurement, quality inspections, etc. But these high-performance technologies usually are less extensively tested, even more so given the working environment (especially radiation exposure) where they will be applied.

This apparent contradiction implies that the adoption of new technologies in the nuclear industry has more difficulties and additional aspects to be considered than those in other common industries.

The objective of ENUSA as a nuclear installation but also as a manufacturing factory, is to manufacture nuclear fuel with the high-quality standards, in strict compliance with the applicable safety rules, and trying to do it as time and cost effective as possible. ENUSA is continuously improving its processes and equipment, especially the most advanced and complex ones.

Focusing on quality inspection in ENUSA's factory, the most advanced equipment is the Automatic Pellet Inspection (API), the rod surface inspection by laser profiler (3DM-Pro), and the Passive Rod Scanner (TESCAN).

Also, some innovation technologies are being applied in those auxiliary or complementary processes around the manufacturing process as control, monitoring, and trace tasks. In this respect, ENUSA is implementing augmented reality to help the operators to use the equipment and as a supporting tool for some manufacturing and maintenance operations.

This article covers the use of deep neural networks in Automatic Pellet Inspection, the rod surface inspection by laser profilometry in 3DMpro equipment, and the use of Augmented Reality in equipment operation.

2. AUTOMATIC PELLET INSPECTION

Pellet visual inspection is the last step of the pellet manufacturing process before introducing them on the fuel rods. The entire pellet production is visually inspected to detect and remove those with defects on its surface.

Prior to the automatic equipment, the inspection was carried out manually by qualified inspectors. The first API equipment was developed by ENUSA in 2000 and a second unit in 2008 based on the first one. In 2016 ENUSA started developing a new generation of vision system (hardware and software), and it was introduced in a third API equipment supplied to Jianzhong Nuclear Fuel Co. in 2019 at Yibin (China).

The functioning of Automatic Pellet Inspection equipment is based on the same concept as that of visual inspection conducted by qualified personnel: capturing the cylindrical surface of the pellets as they rotate between two cylindrical guides to recognize defects on its surface. The image of the lateral surface is captured by a camera and processed by software. Therefore, the inspection is superior to the visual inspection performed by the inspector, as it eliminates the variability linked to human factors such as subjectivity and fatigue; and reduces the risk of radiation exposure [1].

Thus far, ENUSA has upgraded one of its API equipment with the new generation vision system and is working on upgrading the second one. In addition to the inspection system upgrading, one of the innovations in API equipment is image analysis based on Deep Neural Networks (DNN).

The existing methods of image analysis in API equipment rely on conventional image processing techniques, with algorithms that have a limited adaptability to variations in images. This process involves extracting quantifiable attributes from the images. It is likely that more characteristics could be derived from the images and their ROI compared to currently used algorithms, and a thorough investigation into these attributes and their combination could potentially lead to enhanced defect detection and classification. However, the vast array of possible parameters and their various combinations would necessitate considerable time investment, making this approach impractical [1].

The application of deep neural networks in image analysis enhances analysis efficiency as it automates the learning process of the most essential attributes of image and ROI [2] as well as learning of more effective attribute combinations for the detection and classification of defects in pellets. Specifically, convolutional neural networks (CNNs) are particularly effective for processing images [3]. They enable the extraction of features that are increasingly complex and abstract and a combination of these features with various weighting factors. Using a large number of images of pellets with defects (and these defects being properly labelled), DNN based image analysis system is expected to identify and classify them with greater reliability.

Figure 1 shows a simplified scheme of a deep neural network structure for image classification. It can be divided into three stages: CNN, DNN and data flattening. The most significant attributes are identified by convolutional layers with the application of convolutional kernel masks. Following this, max pooling operations are carried out, which serve to reduce convolutional operation results' dimensionality (n × n blocks' the maximum value is extracted). Upon the extraction of all these features, they are arranged into a large array through a process known as flattening, which is then fed into a neural network (fully connected) that produces the classification outcome [1].

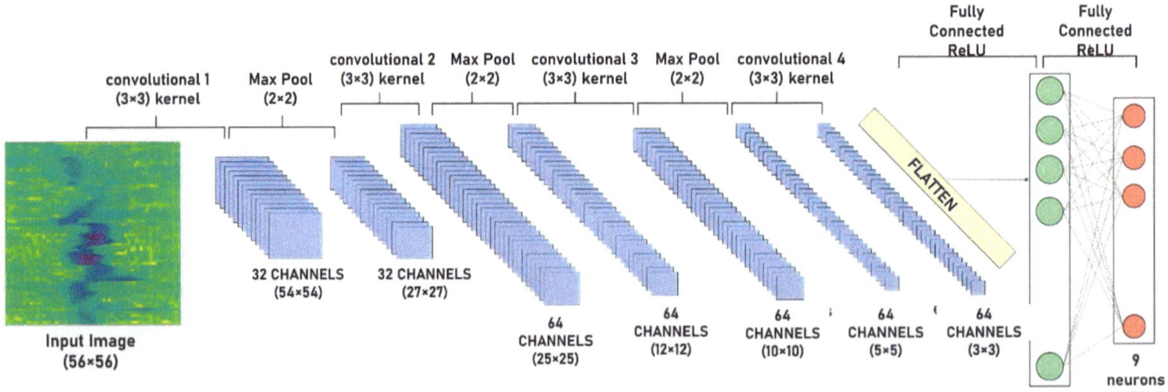

FIG. 1. Basic structure of image classification system based on neural network.

For the training of an image analysis system based on neural networks, it's required to collect a large number of images containing the shapes, features or characteristics to be detected. This image collection could be considered as a database with already known results. The identification and labelling of the results on the images are carried out by experts. Although this labelling process is highly time-consuming it's essential to conduct it meticulously, as the behaviour and results of the inspection are directly influenced by the quality of the labels.

The training is a process of tuning the network by multiple iterations, involving continuous testing of a training set selected from a database. In every iteration, training algorithms adjust the parameters, or weights, of the neural network with the objective of minimizing the error between the result produced and the desired output defined by the labels associated with the data set [1]. The training process is performed for different network configurations (size of input image, different number of layers, number of neurons per layer, interconnections, etc.) to get the optimal network structure.

The availability of a greater amount of data leads to improved results from the neural network, as a large amount of data enables the network to generalize more effectively [1]. An artificial way to increase the data is to generate it in a synthetic way (either creating it from scratch, or by modifying the existing data). This technique is called "data augmentation". ENUSA developed an application to generate synthetic pellet images with defects in two ways: taking the complete image and modifying it (brightness, contrast, flipping, rotating, etc.); or generating a new image by taking the image of a pellet without defects and adding defects extracted from other images.

The results of more significant steps in the development process are indicated below. The result is given by its success rate, calculated by the comparison between the expected result (label) and the obtained result (system classification). The inspection system is able to classify the image in the following defect types: cracks, branch cracks, material losses (end-chamfer and lateral side), pits, end-capping, and no defect (input image is discarded as a defect).

The success level of the first architecture that obtained a relatively reliable result was 75.6%.

Second architecture was obtained by adding the location and size of the defect as network input. The location and size are significant information for defect classification. By adding this information as an input of the DNN, the success rate improved up to 81.3%.

Using data augmentation to generate a bunch of artificial images (30000 samples per defect type, 500000 defects in total), a success rate of 90.08% was achieved.

The following Figure 2 shows the confusion matrix. It represents the success rate in a graphically.

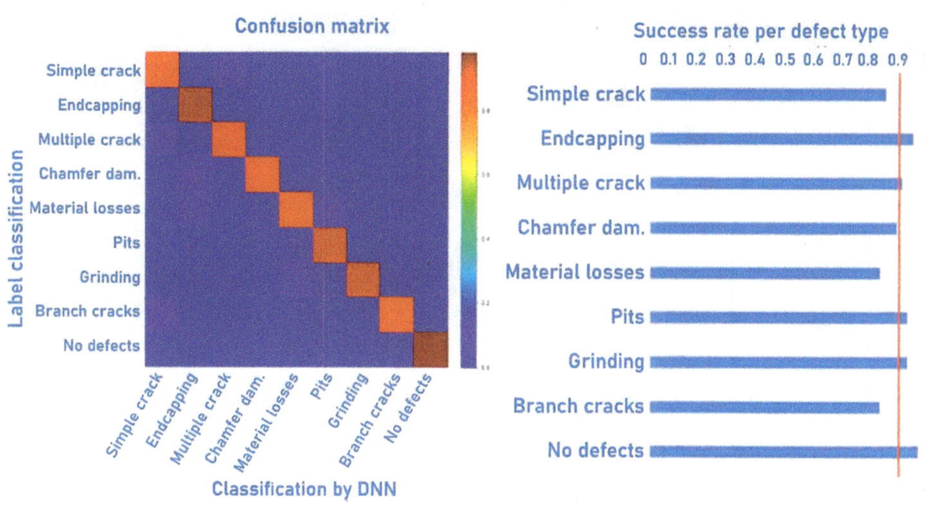

FIG. 2. Confusion matrix obtained by the final network architecture.

The final assessment indicates that inspection of UO_2 pellets is viable using a deep neural networks based system and produces promising results. The primary benefits of employing a system for pellet image analysis that relies on deep neural networks include enhanced reliability in classification and detection, as well as reduced sensibility to changes in image conditions [1].

After the development of the pellet image analysis based on DNNs, ENUSA is working on the integration of API's inspection software in order to test it on real conditions. This will allow the comparison of the system based on DNNs with current inspection systems with the ultimate goal of using it on API equipment.

3. ROD SURFACE INSPECTION BY LASER PROFILOMETRY (3DMPRO)

3DMPro is an inspection equipment to detect defects on the fuel rod surface, installed at the exit of a passive scanner. It's formed by four individual laser profilers, in a cross configuration around the path line through which the rods exit from the scanner.

A profilometry consists of a laser projector and a camera located at a certain angle each other. The laser projects a line on a surface, and the camera gets the variation or deformation of the line, in 2 dimensions, depending on the shape of the surface where it's pointing, and can deduce the shape in 3 dimensions of the surface.

It delivers high resolution distance measurements using laser triangulation techniques. Its configuration makes it possible to cover all the rod circumference perimeter, and with a high capture frequency, it's possible to get a complete map of the rod surface with high resolution to detect defects on it by computer vision. Maximum surface speed would be 200 mm/s, but the scanner works at 90 mm/s.

Defects are measured in depth, length, width and depth, and can be classified as pits, nicks, scratches, and fissures. The maximum allowable defect depth is 25 microns.

ENUSA has also developed manual rod surface equipment based on laser profilometry. It is similar to the automatic 3DMPro equipment, but it only has one profilometer mounted on a desktop support that can be located faced to the FA or over a single rod in an inspection table. It has similar characteristics to automatic equipment, but it's more oriented to defect characterization in more controlled conditions. Both equipment (automatic and manual) is patented by ENUSA. The following Figure 3 shows automatic 3DMPro at the exit of the scanner, and the manual 3DMPro equipment for inspecting the rods already inserted in the FA.

FIG. 3. 3DMPro equipment. Automatic (left) and manual (right).

The following Figure 4 shows the basic architecture of 3DMpro equipment:

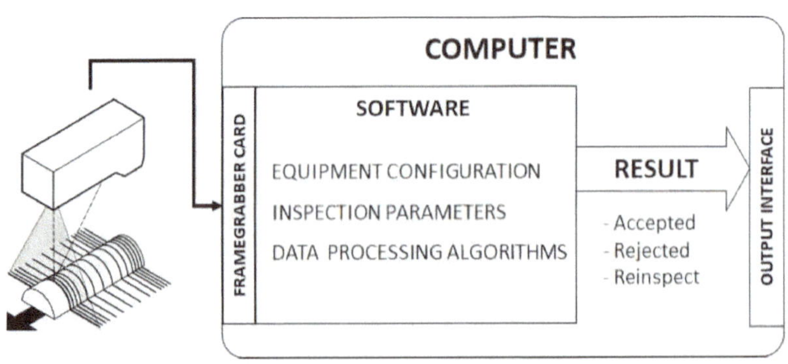

FIG. 4. Basic architecture of 3DMPro inspection system.

According to the specification of the profilometers itself their repeatability under static (ideal) conditions is about 0.2 µm in depth. But once they are installed on the 3DMPro equipment and based on several tests in real factory conditions, the repeatability of the automatic equipment is 10 µm in depth and 100 µm in width and length;

and the repeatability of the manual equipment is 5 μm in depth and 50 μm in width and length. These tests were performed using a standard tube with different characterized defects made on it. Also, during these tests, the correct defect detection and analysis on those areas where the field of view of two profilometers overlapped were checked.

Currently, ENUSA is developing the software for the automatic rod inspection. It will include the algorithms for processing the data obtained from the profilometers, as well as other additional tools for application configuration, equipment adjusting, user management, traceability, activity log, etc.

Thanks to the compact design of 3DMPro equipment, it could be installed on other points of the tubes and fuel rods production line for the inspection of its surface.

Also, thanks to its accuracy, it could be applied to the inspection of other parts and surfaces. It would require previous tests and additional development to confirm the feasibility of its usage on the specific application, and some additional development and adjustments.

4. AUGMENTED REALITY FOR EQUIPMENT OPERATION

Augmented Reality is increasing its presence in the industry: Computer-aided Design (CAD), remote support, distance collaborative tools, training, interactive guidance, etc. ENUSA is also adopting this technology, mainly on equipment operation as training, interactive operation and guidance, and diagnosis and maintenance tasks. This technology is not used for inspection, but it can improve the operation of the equipment in the production line (manufacturing and inspection equipment). So, it can improve the global process and the final product quality.

Augmented reality consists of the projection of images, data, and virtual objects in the user's field of view to overlay them in the real world. Unlike virtual reality, where the user has a screen in front of his eyes and he is in the virtual world, in an augmented reality application the user can still see the real world with more information and objects added to it.

An augmented reality system has two parts: the glasses (the hardware) and, equally important, the application (the software).

ENUSA has developed some test applications to test this technology and acquire the knowledge. The following Figure 5 shows a screenshot of one of these applications as the user sees it.

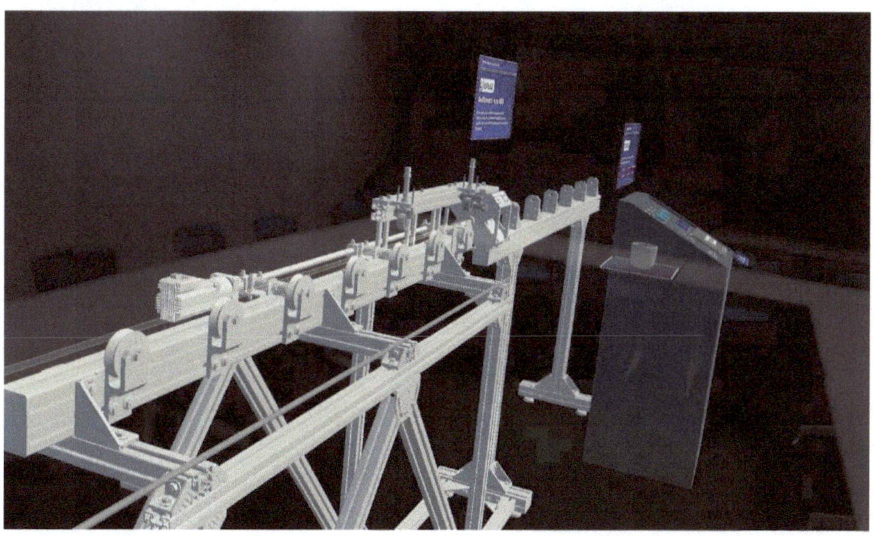

FIG. 5. Virtual model of laser profilometer inspection prototype in an Augmented Reality application.

Now ENUSA is developing a complete application in production equipment to assist the operator in a consumable change operation that needs to be done twice per shift. This operation has been chosen because it covers all the features of an industrial application using Augmented Reality:

- Continuous and real-time communication with equipment automation;
- Spatial location using complete equipment virtual model as a reference;
- Object recognition and presence detection of some specific parts;
- Interactive operator guidance including access to procedure documentation;

- Trace of the operation and its result to factory Manufacturing Execution System.

This application will be also used as a basis for future ones for various equipment and processes, the development of which will be faster thanks to the acquired knowledge.

5. CONCLUSIONS

ENUSA, in its constant commitment to research new technologies, continues developing and improving its equipment and processes by applying new and innovative solutions. Automatic Pellet Inspection, 3DMPro, and the introduction of Augmented Reality are real examples of these developments. Not only process improvement, but also better quality, safety, and effective cost are the main ENUSA goals in the manufacturing investment area.

REFERENCES

[1] VERDEJO, D. et al., Application of deep neural networks in automatic visual inspection of UO2 pellets (2022)

[2] LECUN, Y., et al., Deep Learning. Nature 521 (2015).

[3] RAWAT, W., WANG, Z., Deep convolutional neural networks for image classification: A comprehensive review, Neural computation 29.9 (2017).

ID#19

PASSIVE SCANNER FOR URANIUM AND GADOLINIUM FUEL RODS INSPECTION

J. LAFUENTE SEVILLANO, A. AHUFINGER MUÑOZ
Quality Engineering,
ENUSA Industrias Avanzadas S.A., S.M.E. (Grupo SEPI),
Juzbado, Spain

Abstract

The passive scanner for the inspection of fuel rods, manufactured by Tecnatom, which was launched this year at the ENUSA FA Factory (Juzbado), is a new inspection equipment based mainly on gamma spectrometry technology and eddy current system equipment, capable of measuring various characteristics of UO_2 and UO_2–Gd_2O_3 fuel rods. Despite being a complex device, its operating principle is simple. This equipment passes fuel rods at a uniform speed through the inspection lines and generates signals that are analysed by automatic algorithms to obtain measurements and search for defects.

The main reasons for investing in a new passive scanner over an active scanner range from improved detection capabilities to impact on facility and worker safety, as well as environmental impact and financial impact.

1. INTRODUCTION

For 35 years, the ENUSA FA Factory (Juzbado) has used an active scanner to inspect 100% of UO_2 fuel rods production (95% of total production). This scanner is made up of inspection equipment based on gamma spectrometry and eddy current equipment.

The designation of 'active' refers to a type of technology which requires a neutron source that increases the activity of gamma radiation of the UO_2 pellets that are inside the fuel rod to obtain a higher enrichment signal. This active scanner, whose gamma detection block has 12 NaI detectors, incorporates a series of ^{252}Cf neutron sources that, due to the decay of this powerful neutron-emitting isotope (half-life of 2.6 years), needs to be periodically reloaded.

In addition, ENUSA Factory at Juzbado also has an operational passive scanner (which does not have a radioactive source for gamma detection) that is used to inspect all fuel rods of UO_2–Gd_2O_3 product which corresponds to approximately 5% of the annual production, as its production capacity is limited by the number of gamma detectors and therefore its inspection speed.

This article describes ENUSA's reasons for investing in a new Double Line Passive Scanner (DLPS) (Figure 1) to inspect 100% of the production, that is, both UO_2 and UO_2–Gd_2O_3 fuel rods. Furthermore, the characteristics of every individual equipment which makes up the whole passive gamma scanner are detailed, along with its functionalities that will make it possible to provide a higher added value to the manufactured product required by our customers.

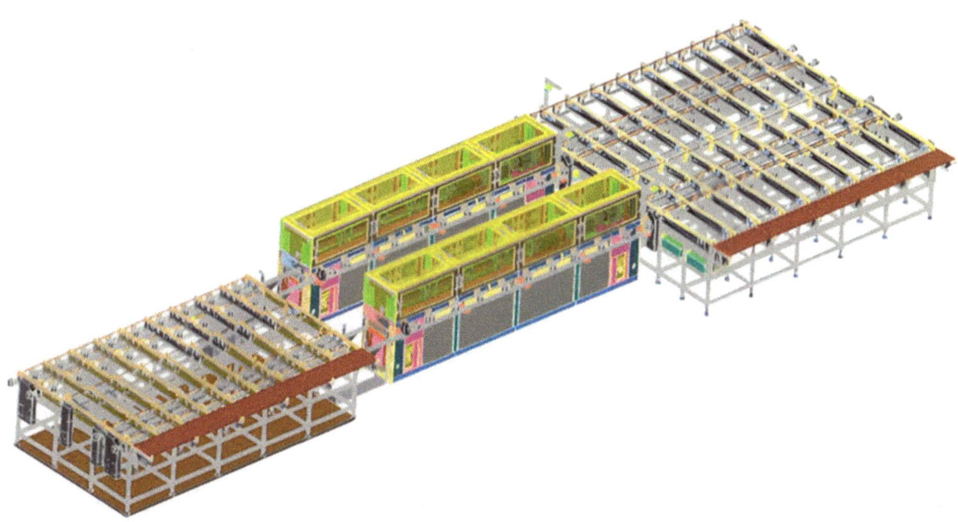

FIG. 1. Layout of DLPS (courtesy of Tecnatom).

2. PASSIVE SCANNER VS. ACTIVE SCANNER

ENUSA is an organization committed to continuous improvement from a quality, production, development, and safety standpoint. Thanks to these principles, the reasons ENUSA has decided to invest in a new passive gamma scanner for the inspection of fuel rods can be clearly identified. This scanner has been manufactured by Tecnatom and launched jointly by both collaborating companies.

Among all the improvements offered by this technological change, three of them stand out:

- Removal of ^{252}Cf neutron sources:
  - ^{252}Cf is a powerful neutron emitting radioactive isotope. This poses a risk to people and all living beings, so it is essential to exercise extreme caution when installing these sources;
  - Because of the ^{252}Cf sources, after UO_2 fuel rods are inspected in the active scanner, the fuel rods' gamma radiation levels increase, so these have to wait for a few minutes before continuing the inspection process to minimize the doses received by workers (ALARA principle[2]);
  - Due to the decay in their activity, ^{252}Cf sources need to be reloaded periodically (every 1 or 2 years) to maintain an optimal enrichment signal for inspection;
  - This sometimes means modifying the inspection speed by having to recalibrate certain scanner parameters, which is a handicap for gamma scanner productivity and stability;
  - ^{252}Cf is a substance that is obtained artificially in laboratories and has a very high price being, if not the most, one of the most expensive chemicals on the planet.
- Improved detection capacity, equipment stability and re-inspection rate:
  - The detection capacity supposes an improvement in the repeatability of the results observed in the qualifications;
  - The stability of the Gadolinium sensor is a significant improvement over what currently exists in the old passive scanner;
  - The re-inspection rate has been reduced, observing in some cases that it is less than 1% in this DLPS compared to the 10–20% that was usually observed in the old equipment for the same product design.
- Updating of the equipment at the operation level, automation, hardware, and software. As it consists of brand new equipment, it has significant improvements over other scanners. Such as speed monitoring, automatic correction of fresh uranium fuel, detector calibration tools, automatic recirculation of suspicious fuel rods, offline analysis tools, new analysis algorithms, current operating systems, etc.

Table 1 summarizes the most important individual equipment and elements of each scanner currently in the factory.

TABLE 1. EQUIPMENT AND ELEMENTS OF THE GAMMA SCANNERS AT JUZBADO FACTORY

Characteristic	**Active Scanner (obsolete)**	**Passive Scanner**	**DLPS (ENUSA-Tecnatom)** ***(per inspection line)***
^{252}Cf Radioactive Sources	✓	-	-
Eddy Current	✓	✓	✓
Densitometer	✓	✓	✓
Gadolinium Sensor	-	✓	✓
Gamma Module	12 detectors	32 detectors	126 detectors
Inspection Speed Monitoring System	-	-	✓
Inspection Speed	120-180 mm/s	20-30 mm/s	60-100 mm/s
Production Percentage	95%	5%	100%

[2] As Low As Reasonably Achievable

3. DOUBLE LINE PASSIVE SCANNER INSPECTION EQUIPMENT

This DLPS for the inspection of UO_2 and UO_2–Gd_2O_3 fuel rods is the result of bringing together all the necessary components so that it is capable of absorbing the inspection capacity of ENUSA's two former gamma scanners.

In Figure 2 distribution of the different inspection equipment of each inspection line of the DLPS is represented.

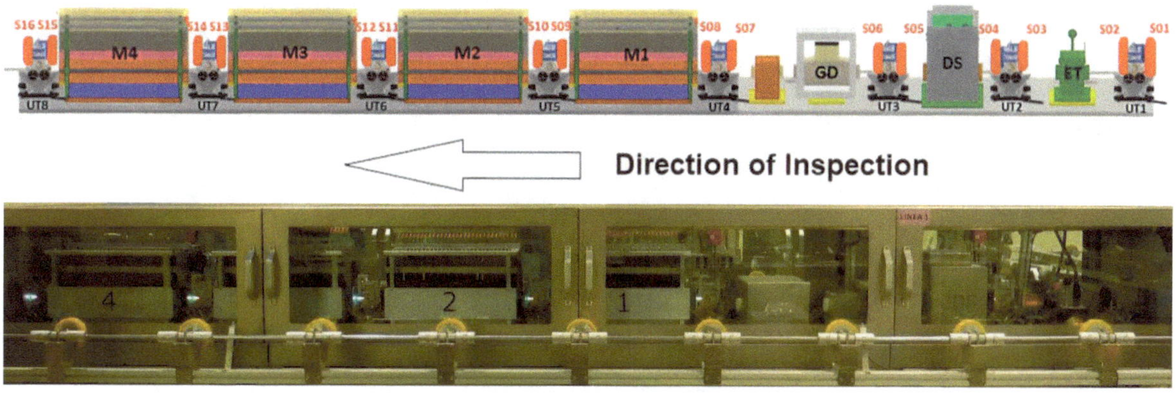

FIG. 2. Schematic view (top) and image (bottom) of the DLPS inspection module.

The following sections describe all the inspection equipment which make up the DLPS.

3.1 Eddy current equipment

Equipment based on eddy current technology for the detection of surface defects in the fuel tube such as notches, cracks, etc. It has multi-frequency electronic equipment connected to an encircling coil through which all the fuel rods pass.

Figure 3 represents an image of a tube standard with defects and its corresponding signal in the DLPS Software.

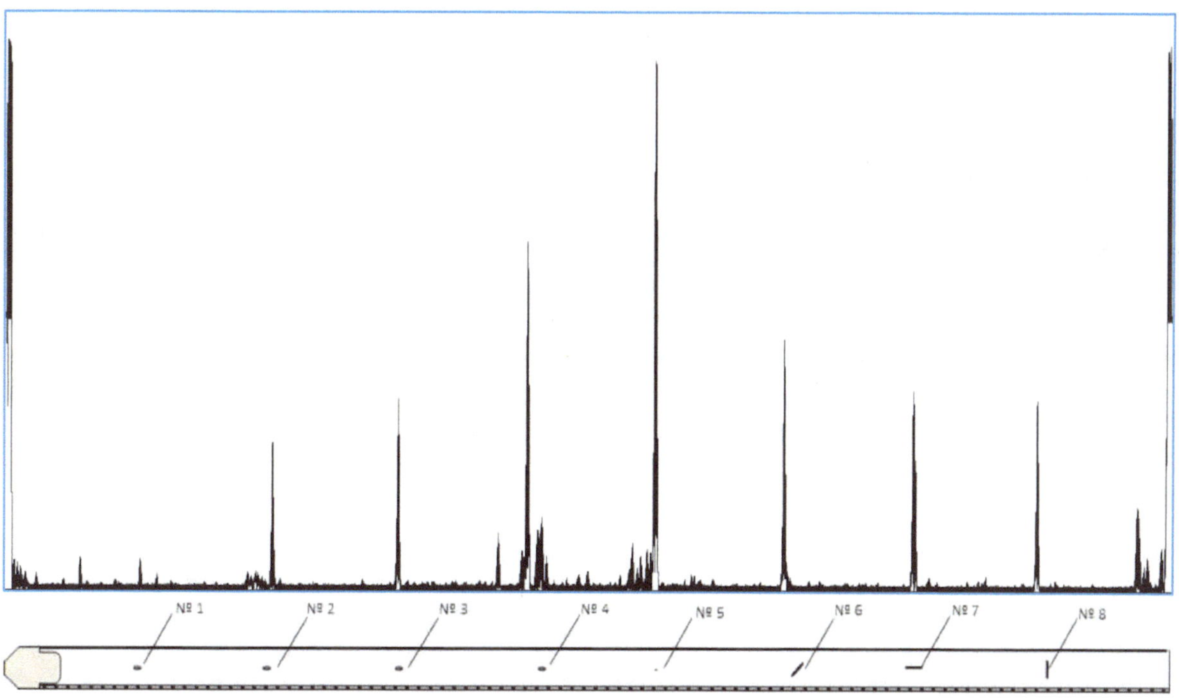

FIG. 3. Sketch of tube standard with defects (bottom) and corresponding signal from eddy current equipment (top).

3.2 Densitometer

Equipment based on gamma ray densitometry that has two sources of ^{241}Am gamma radiation (half-life 432 years) oriented perpendicularly to each other and facing a Bismuth Germanate gamma detector. The resulting signal is the average of the two individual signals. The signal is maximum when there is no opposing material (fuel rod) between the sources and Bismuth Germanate detectors and will decrease according to the density of the fuel rod and its overall density.

This equipment measures mechanical characteristics of the fuel rod such as rod length, active column length, plenum length, spring compression zone length (exclusive GNF2[3] designs), spring presence and detection of gaps in the active column.

Figure 4 represents an image of a PWR technology fuel rod standard with artificial gaps and its corresponding signal in the DLPS Software.

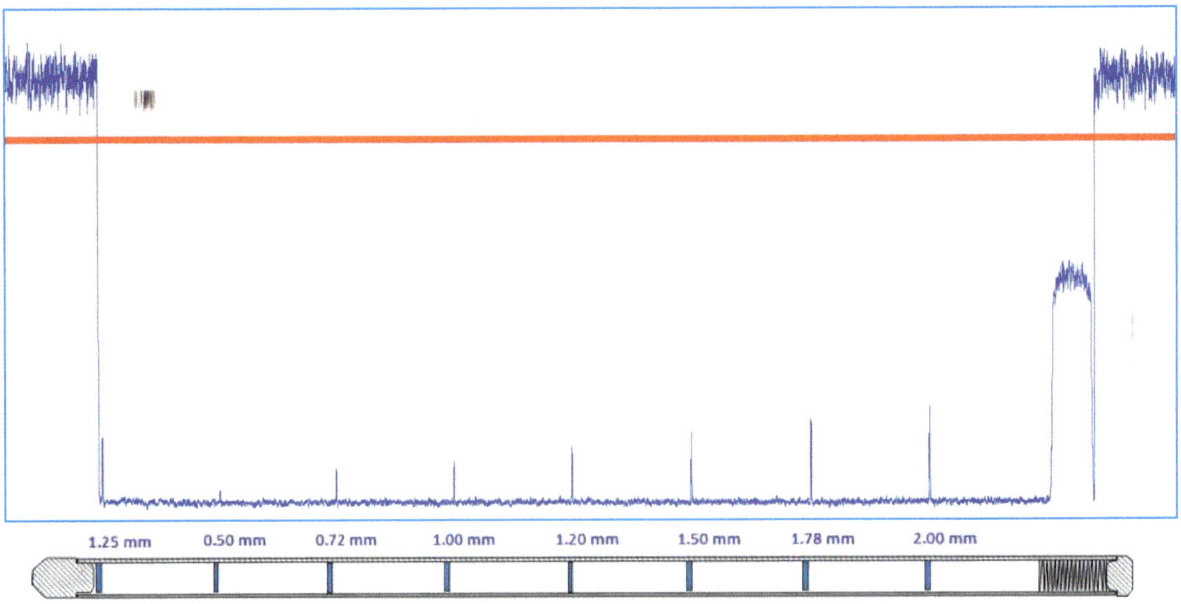

FIG. 4. Sketch of a fuel rod standard with gaps in mm (bottom) and corresponding signal from the densitometer (top).

3.3 Gadolinium Sensor

The function of the Eddy current equipment is to measure and detect variations in the content of Gd_2O_3 in pellets. Since this substance has a high magnetic susceptibility compared to UO_2, these variations in Gd_2O_3 will be reflected in the signal displayed by the sensor. The probe consists of an encircling coil with a permanent magnet to eliminate possible noise variations in the signal caused by the Gadolinium magnetic properties.

This equipment measures average Gadolinium content, Gadolinium zone length, Gadolinium off-specification pellets, and pellet lack of material. In addition, there is a complementary encircling coil but without a magnet which is capable of differentiating between a positive off-specification pellet in Gadolinium content or the presence of metallic particles inside the pellets by comparing the two signals.

Figure 5 represents an image of a PWR technology fuel rod standard with Gadolinium off-specification pellets and its corresponding signal in the DLPS Software.

---

[3] BWR Nuclear Fuel designed by Global Nuclear Fuel

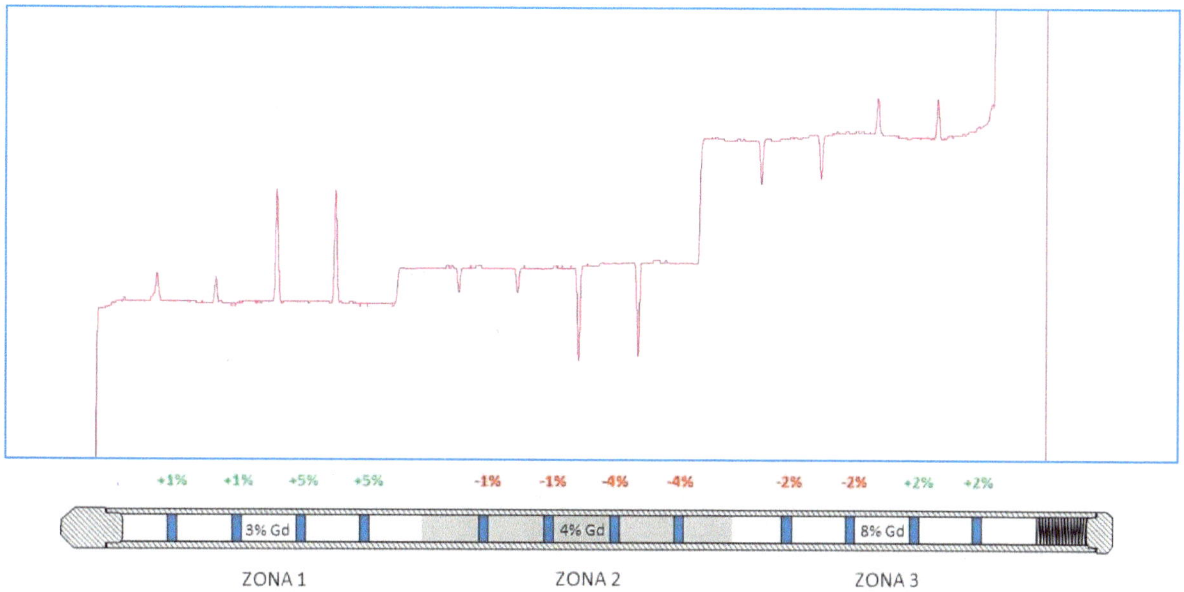

FIG. 5. Sketch of fuel rod standard with Gadolinium off-specification pellets in absolute % deviation (bottom) and a corresponding signal from the Gadolinium sensor (top).

3.4 Gamma Detection Modules (M1-M4)

The Gamma Scan rod inspection consists of four modules, each one made up of 32 Bismuth Germanate gamma detectors, with the exception of the M1 module that contains only 30 detectors, since the other two slots correspond to those of the densitometer, thus making up 126 gamma detectors per line.

The gamma detection block constitutes a global enrichment signal composed of each of the signals from the 126 gamma detectors. This high number of detectors compensates for the absence of a ^{252}Cf neutron source which otherwise would activate the pellet column increasing the enrichment signal level counts. This is why, due to the number of signals that make up a single joint signal, the synchronization of all of them is critical and, consequently, the uniformity of the fuel rod inspection speed.

The gamma detection system is capable of measuring pellet active column length, enrichment zone lengths, average enrichment percentage, and detecting off-specification pellets in enrichment.

Figure 6 represents an image of a PWR technology calibration fuel rod standard with different enrichment zones and its corresponding signal in the DLPS. These types of standards are used to calibrate the equipment both in inspection speed (length) and in enrichment (counts/second to average enrichment percentage).

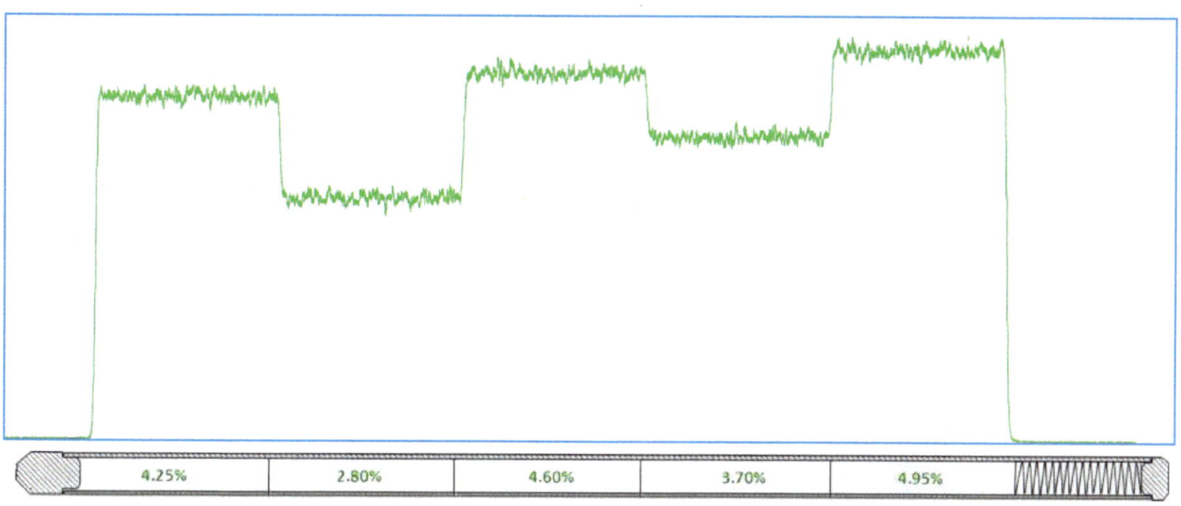

FIG. 6. Sketch of fuel rod standard with zones of different enrichment % (bottom) and corresponding signal of the gamma signal (top).

Figure 7 represents an image of a fuel rod standard of PWR technology with off-specification pellets in enrichment and its corresponding signal in the DLPS.

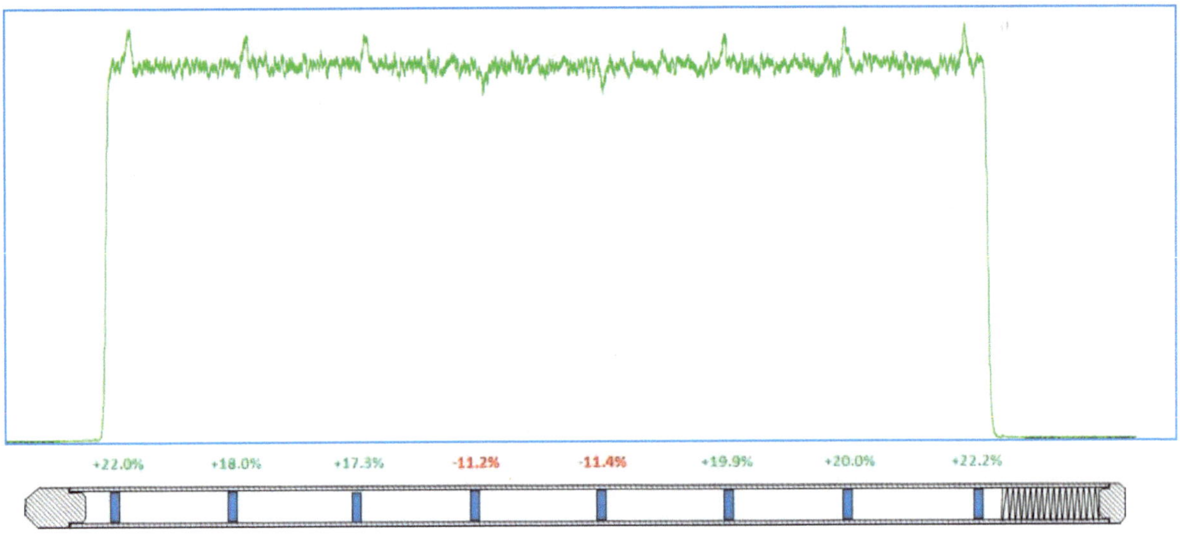

FIG. 7. Sketch of fuel rod standard with off-specification pellets enrichment in relative % deviation (bottom) and corresponding signal of the gamma signal (top).

Finally, it is to be noted that all the signals generated by the DLPS are analysed using automatic algorithms right after each fuel rod passes through the inspection line. In addition, each inspection line contains about 300 editable parameters, which are dependent on design, detection capabilities, and signal analysis.

3.5 Laser barrier sensors for speed monitoring and step motors.

16 barrier sensors are installed to measure and monitor the fuel rod inspection speed (controlled by eight step motors) which is one of the critical inspection parameters.

Laser barrier sensors have a fixed position (in mm) all across the inspection line, and they record time stamps as each fuel rod passes through them to calculate the inspection speed of the top end and bottom end plug. Variability of position (mm) vs time stamps (sec) make up the inspection speed uniformity.

4. CONCLUSIONS

In light of the above, the main advantages that the incorporation of the new DLPS to the fuel rod inspection line have been identified for ENUSA and its clients are listed below:

- Impact on the safety of the facility and workers. Environmental and economic impact. All this is favoured thanks to the elimination of the neutron sources of ^{252}Cf;
- Improved scanner performance in terms of detection/measurement, repeatability, and efficiency;
- Incorporation of inspection control elements and other improvements compared to former scanners.

All this leads to an important added value to the product manufactured from the point of view of ENUSA and the customer.

Finally, to obtain a basic overview of the different equipment that makes up this DLPS and its performance, a brief description and demonstration of each of them has been presented.

TECHNICAL SESSION V: EXPERIENCE AND PROSPECTIVES FOR FUEL FABRICATION

ID#21
ARGENTINE EXPERIENCE

L. ALVAREZ
Fuel Engineering Department,
National Commission on Atomic Energy (CNEA),
Buenos Aires, Argentina

Abstract

The present contribution details relevant aspects in relation to the development and application of the technology associated with fuel elements for NPPs in Argentina.

1. INTRODUCTION

The development and application of technology for nuclear fuels have been carried out in Argentina for more than 40 years based on a strategic conception regarding the importance of nuclear technology and its application in developing countries. One of the clearest inspirers of these ideas was Professor Jorge Sabato (1924-1983), one of the most relevant leaders in Argentine nuclear development, who stated that an NPP is more than a kilowatt-hour factory; it is the key to the technological and industrial transformation of the country.

2. NUCLEAR ENERGY IN ARGENTINA

There are three NPPs in operation in Argentina:

- Atucha 1 (CNA-1);
- Embalse (CNE);
- Atucha 2 (CNA-2).

Their combined contribution at full capacity to the Argentine production of electricity is about 9%.

The three NPPs have PHWRs cooled and moderated by heavy water and with on-power refuelling and fuel reshuffling.

The following paragraphs present some data of the NPPs like the year of starting operation and the current gross electrical power. Different images of these plants are also provided.

2.1 Atucha 1 NPP (Vertical PHWR) – 1974 – 362 MW

The following Figure 1 shows photographs of the construction and shipment operation of the pressure vessel manufactured in Europe. The transportation of this main component, of large size and weight, at that time, required to solve important logistical challenges (for example reinforcement of bridges and adaptation of electrical lines that were crossing the roads).

(a) (b)

FIG. 1. (a) Construction of Atucha 1 and (b) shipment of its pressure vessel.

2.2 Embalse NPP (CANDU Type – Horizontal Pressure Tubes) – First Operation 1984 – Refurbishment and Life Extension 2019 – 656 MW

The Figure 2 (a) shows part of the initial construction related to the positioning of the cooling channels with their pressure tubes. The Figures 2 (b) and 2 (c) show the appearance of the plant already in operation and the device for transporting to the plant one of the four steam generators that replaced the old ones as part of the life extension works. All the new steam generators were manufactured in Argentina.

(a)

(b)

(c)

FIG. 2. (a) Construction of the Embalse NPP, (b) Current aspect of the NPP, and (c) Transportation of the new Steam Generators.

2.3 Atucha 2 NPP (Vertical PHWR) – 2014 – 745 MW

The following photographs (Figure 3) are related to the Atucha 2 NPP. Figure 3 (a) shows work on the top of the pressure vessel for the placement of the FAs of the first core in the cooling channels and the Figure 3 (b) shows the Atucha 1 and Atucha 2 NPPs locations respectively on the shore of the Parana River.

(a)

(b)

FIG.3.Atucha 2 NPP.

The Embalse and Atucha 2 NPPs are fuelled with natural uranium. Although the Atucha 1 plant was also initially designed to use natural uranium, currently, and since the year 2000, it has been using FAs with uranium slightly enriched to 0.85% in ^{235}U.

The entirety of the FAs for the three NPPs is produced within Argentina. Their designs have progressively evolved over the years, influenced by operational insights, advancements in fabrication, and the demands of technical and economic requirements. The main characteristics of the three NPPs and their FAs are reported in more detail in other contributions also presented at Technical Meeting held in 2021 (ID#11 [1] and ID#9 [2]).

3. MAIN FIELDS OF DEVELOPMENT

Along with the construction and operation of the NPPs the nuclear fuel technology was also developing and reaching a great degree of maturity.

The main topics on which this development took place were:

- Material Sciences and Metallurgy, including the Irradiation Effects during the nuclear service;
- Product Engineering including the simulation with computer codes of the fuel behaviour during the irradiation process and also during the storage of the spent fuels;
- Manufacturing Processes Development;
- Non-Destructive Testing and other QC Techniques;
- QA;
- Post-Irradiation Examination (Pool Side Inspection and Hot Cell Techniques).

Several of these developments led to their extension to non-nuclear applications and the corresponding spin-off of knowledge and services, which contributed to strengthening the role of the CNEA as a knowledge and technology advisor and as a recognized provider of services for the development of the non-nuclear local industry.

4. EVOLUTION OF NUCLEAR FUEL TECHNOLOGY

Some relevant milestones related to the development of fuel technology for NPPs in Argentina are detailed in the following paragraphs.

1982

- Reception of the Technology transferred by SIEMENS for the fabrication of the FAs for the Atucha 1 NPP;
- Development of the local product engineering for the manufacturing of the Atucha 1 Fuel;
- Beginning of the domestic supply of Atucha 1 FAs to replace the German (RBU) supply.

1984

- Local development of the fuel manufacturing technology for Embalse FAs (CANDU Type);
- Local development of the product engineering and qualification of fabrication processes;
- Beginning of the fuel manufacturing to gradually replace the FAs of the first core supplied by Canada.

1995–2000

- Development of the local engineering for the replacement of the fuel at Atucha 1. Natural Uranium was replaced by SEU. Fuel engineering activities included:
  - The development of drawings and specifications for the new situation;
  - The preparation of the licensing documents;
  - The qualification of the manufacturing process to obtain SEU UO_2 powder obtained by blending natural UO_2 powder and enriched UO_2 powder. The process to obtain SEU fuel pellets from the above-mentioned blending was also set up and qualified.

2009–2014

- Adaptation of the Atucha 2 manufacturing technology to the local industry conditions and local development of fabrication processes;
- Development of product engineering based on the fuel engineering supplied by KWU and in the local experience and possibilities.

— Domestic manufacturing of the complete first core for the Atucha 2 beginning of operation;
— Development of transportation requirements and equipment for the delivery of the fuel from the fuel manufacturing plant to the NPP site.

2016

— Local development of the fuel engineering for the CAREM Project;
— Local development of the manufacturing technology for the CAREM FAs;
— Beginning of the domestic fabrication of the first core for the CAREM reactor.

5. ARGENTINE NUCLEAR FUEL NETWORK

The main organizations and tasks that are involved in Argentina in the development and application of nuclear technology in the field of FAs for power reactors are:

National Commission on Atomic Energy (CNEA)

— Development and fuel engineering, design, product engineering, and modelling activities;
— Materials Testing Laboratories and Research Facilities

CONUAR-FAE and DIOXITEK

— UO_2 powder, cladding tubes, components, and FAs manufacturing processes development;
— Fuel manufacturing and transportation.

NASA

— Nuclear Power Plants operation;
— Detection of fuel failures;
— Standard Post Irradiation Examination activities;
— Licensing presentations and relationship with the Nuclear Regulatory Authority.

As previously indicated, nuclear fuel manufacturing activities are carried out mainly at the CONUAR-FAE facilities located at the Ezeiza Atomic Center. Figure 4 below shows a view of the manufacturing complex with both fabrication plants. FAE is mainly involved with the manufacturing of tubes for claddings and other utilizations and CONUAR activities are related to the fabrication of the FAs and their components.

FIG.4. FAE and CONUAR Fuel Manufacturing Complex.

In addition to the above-mentioned organizations, the ARN (Nuclear Regulatory Authority of Argentina) also plays an important role. Its primary function is to regulate and supervise nuclear activities, particularly in the field of safeguards and non-proliferation, protection and physical security, and radiological and nuclear safety.

6. ACCUMULATED EXPERIENCE

Manufacturing of nuclear FAs

The following table shows the number of FAs of each type manufactured since the beginning of CONUAR-FAE's operation until December 2020. These figures in Table 1 are a clear indicator of the extensive experience accumulated in almost forty years of fuel manufacturing in Argentina.

TABLE 1. NUMBER OF FUEL ASSEMBLIES/FUEL RODS MANUFACTURED FOR THE DIFFERENT NPPS

NPP	FAs manufactured (up to December 2020)
Atucha 1	9191
Embalse (CANDU)	129632
Atucha 2	2470
CAREM	1160 Fuel Rods

7. MAIN DEVELOPMENT TOPICS ON NUCLEAR FUEL TECHNOLOGY

The main topics of interest on which activities are currently being carried out in Argentina in relation to fuel technology are the following:

- Improvement of Fuel Pellets Mechanical Resistance (all types of fuels):
  - Reconsideration of the type of internal and external defects included in the fuel pellet specifications, their potential origins, and their relevance during the nuclear service and transportation;
  - New standard tests are under development and qualification.
- Beryllium Brazing replacement (Embalse-CANDU fuel):
  - Based on environmental considerations and health risk reduction one of the important activities is related to the replacement of the use of Be for the brazing of appendages at the Embalse fuels;
  - Alternative materials for the brazing process and also new processes are under consideration.
- Improvement of Fuel Resistance to Power Ramps - The main objectives are to reduce the risk of failures associated with pellet cladding interaction and to provide more flexibility to the operation of the NPP. Some of the activities are related with:
  - New Internal Coatings;
  - Improved cladding materials;
  - Improved cladding Manufacturing Process.
- Conversion of Atucha 2 to SEU – Based on lessons learned from the experience in Atucha 1 regarding the economy of the fuel contribution to the cost of generation, the reduction in the consumption of FAs, the reduction in the utilization of the loading machine, and the reduction of the volume of spent fuel, a similar programme is being planned for Atucha 2;
  - CNEA and NASA already agreed on a roadmap for the conversion which includes engineering document development, endurance tests, and a lead test assemblies irradiation programme.

8. FINAL REMARKS

As presented in this contribution, the implementation of nuclear technology in a country like Argentina represents the opportunity to develop fuel technology in parallel. This technology allows identifying needs for the development and optimization of FAs but also allows generating knowledge and motivating the acquisition of information and equipment that can also be applied for the supply of advice and services to other fields of the industry that may need them.

REFERENCES

[1] BUSSOLINI, A., et al., Alternatives to replace the use of Beryllium for appendages attachment on CANDU Fuel Elements, IAEA TM on Technical Challenges and Advances in Fuel Fabrication, Vienna, Austria, 8-10 November 2021.

[2] MEDINA J. P., et al., Developments and Methodology for the Characterization of the Mechanical Integrity of PHWR Fuel Pellets, IAEA TM on Technical Challenges and Advances in Fuel Fabrication, Vienna, Austria, 8-10 November 2021.

ID#22
NUCLEAR FUEL R&D FOR COMMERCIAL REACTORS IN KAERI

J. H. YANG, D-J. KIM, D-S. KIM, Y-H. KOO
LWR Fuel Technology Division,
KAERI,
Daejeon, Republic of Korea

Abstract

KAERI is a government-funded research institute established in 1959 with the goal of achieving energy self-reliance through nuclear energy. As Korea's only research institute dedicated to nuclear energy, KAERI has played a pivotal role in advancing the country's nuclear fuel technology over the past 60 years. Presently, KAERI is committed to the development of cutting-edge nuclear fuel technologies with improvement in safety and economy. Our endeavours include international cooperation, contributing to the global progress of nuclear fuel technology.

1. INTRODUCTION

A chronological timeline of key milestones related to the construction of nuclear reactors in Korea, as well as the export of nuclear reactors abroad is depicted in Figure 1 [1]. The first milestone in the timeline involves the construction of the TRIGA research reactors. Two research reactors were constructed with the aim of fundamental research, education, and producing radioisotopes. During this phase, nuclear fuel was imported from overseas.

In the 1970s, Korea initiated the construction of commercial power reactors, including both LWRs and HWRs. KAERI has been committed to achieving the localization of nuclear fuel supplied to Korean domestic commercial NPPs. From the late 1970s, KAERI introduced conversion technology and built a UO_2 powder manufacturing facility to support the localization of nuclear fuel for commercial NPPs under construction at that time. In the 1980s KAERI successfully completed the performance verification of domestically developed HWR fuel prototypes with Atomic Energy of Canada Limited and established mass production capabilities. In conjunction with these efforts, KAERI actively participated in the localization of LWR fuel through collaborations with domestic industries and international organizations such as Siemens-KWU and Combustion Engineering (C-E). This contributed to achieving self-reliance in LWR fuel design technology. In 1995, a multi-purpose research reactor HANARO was built. KAERI developed the rod-type U_3Si fuel for HANARO and supplied HANARO fuel.

In the early stages of technology development, KAERI cooperated with overseas institutions. For the HWR and Research Reactor fuels, we cooperated with Atomic Energy of Canada Limited in Canada. In the case of LWR fuels, we cooperated with Siemens-KWU and C-E on the nuclear fuel design.

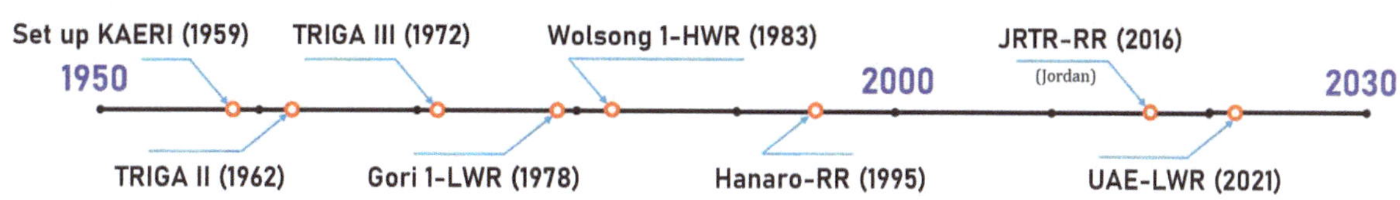

FIG. 1. Chronological timeline of key milestones related to construction of nuclear reactors.

In 1996, responsibilities related to the design and manufacture of nuclear fuel for commercial power reactors shifted to Korean industries. Since then, KAERI's role has evolved into pioneering innovative nuclear fuel technologies for LWRs and supporting the industry in implementing these advancements. KAERI is at the forefront of developing GEN IV nuclear fuel technologies and advanced research reactor fuels.

Figure 2 shows the fuel research topics under study in KAERI. For research reactor fuel, KAERI supplies rod-typed U_3Si fuel for HANARO. The development of plate-type U–Mo fuel is underway for a new domestic research reactor under construction. Furthermore, KAERI is actively involved in the development of plate-type fuel using U_3Si_2 for deployment in overseas research reactors. The unique centrifugal powder manufacturing method developed by KAERI is employed in the production of fuel powders.

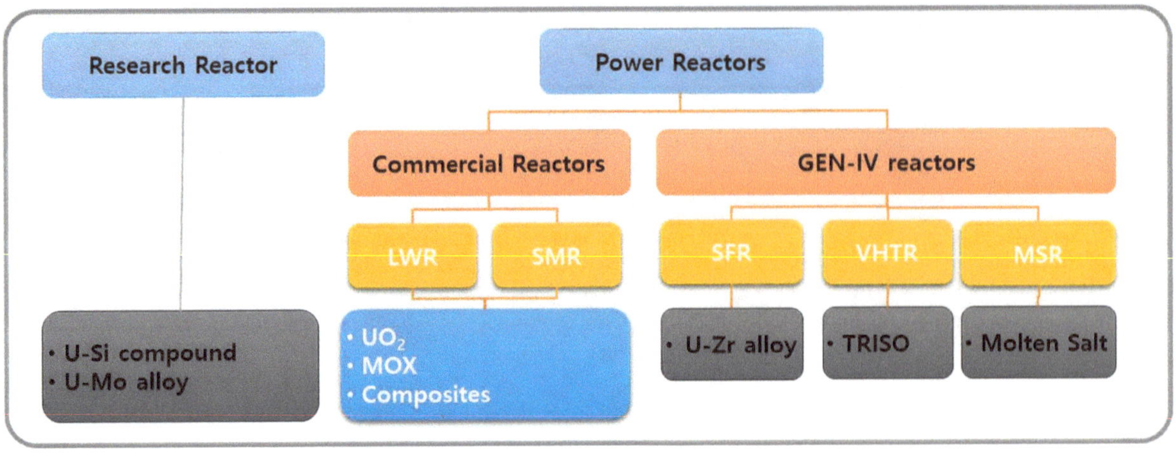

FIG. 2. Fuel research topics under study in KAERI.

Regarding the fuel for GEN IV reactor systems, efforts are in progress to advance the technology of transuranic radioactive waste containing U–Zr alloy fuels, with the aim of minimizing the volume of highly radioactive elements by burning them in Sodium Fast Reactor. Research on TRISO fuel for gas cooled reactors includes the use of non-oxide uranium nitride (UN) kernels and zirconium carbide (ZrC) coated layer. KAERI is also leveraging its expertise in pyro-processing technology with molten salt to develop chloride-based fuels for molten salt reactors. The next chapter provides a more detailed overview of the R&D status of LWR fuels.

2. LWR FUEL R&D IN KAERI

With the development of nuclear technology and changes in technological demand, research topics on fuels for LWR have changed gradually. The research conducted in KAERI so far can be divided into three stages according to the purpose and level of technology as shown in Figure 3. In the first phase, KAERI focused on technology development to improve the quality of existing nuclear fuel pellets. A manufacturing process for UO_2 pellets has been developed, utilizing U_3O_8 powder obtained through low-temperature oxidation, along with a micro-doping of alumina [2]. This process enhances the grain size and thermal stability of the resulting UO_2 pellets. The manufacturing process of a gadolinia pellet with excellent quality in terms of chemical homogeneity and pellet microstructure was also developed. These technologies were transferred to industry and utilized for quality improvement and localization. In the next step, research mainly focused on innovative concepts to improve the economics of nuclear energy. KAERI has developed annular pellet technology for dual cooled fuel that can enable power uprate [3]. A duplex burnable pellet with different poison compositions in the outer and inner zones was developed for long fuel cycle operation.

The paradigm of nuclear fuel research has changed significantly since the occurrence of an accident involving hydrogen explosion. Safety is more emphasized these days and our recent R&D efforts are focused on the development of ATF with new material concepts, which has improved safety [4-6]. Initiatives on LEU+ fuel research, which is to extend burnup and enrichment beyond legacy limits, have commenced to diminish the amount of spent fuel and enhance economic viability. The soluble boron-free core is a major design requirement for advanced LWRs such as SMRs. Advanced burnable absorber fuels and control rod materials are being concurrently developed in conjunction with LEU+ fuel and boron-free-core to regulate the anticipated excess-reactivity. To expedite technological advancement, advanced technologies such as composite design and AM are actively being incorporated.

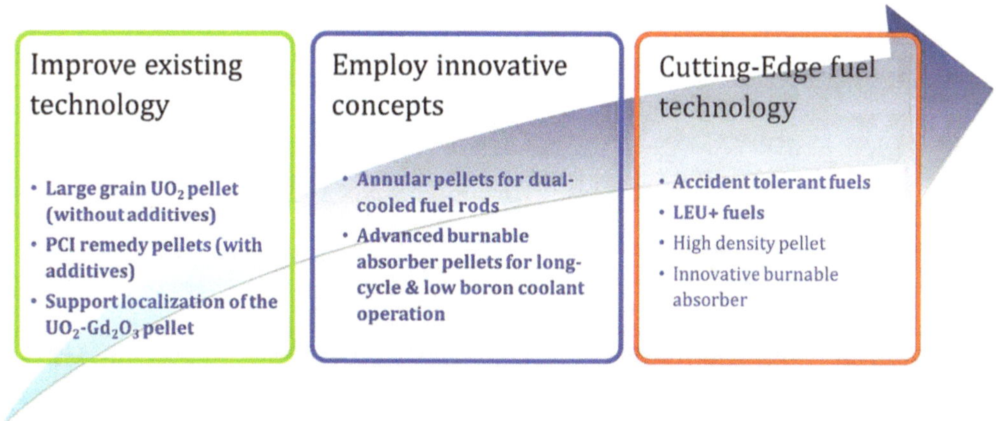

FIG. 3. Progress in LWR fuel research in KAERI.

3. CHALLENGES IN DEPLOYMENT OF ADVANCED LWR FUEL TECHNOLOGY

In order to demonstrate safety and to commercialize nuclear fuel of new innovative concepts, it is necessary not only to develop manufacturing technology but also to conduct long-term irradiation tests in reactor-relevant environments. The reliability, performance, and safety of the fuel are primarily supported by experimental data obtained through post irradiation examination of irradiated fuels under normal and accidental conditions. Additional data to support the licensing of post discharge spent fuel management and disposal operation is also needed.

Innovative types of ATF or fuel with LEU from 5% to 10% (LEU+) will require more R&D and infrastructure investments to be deployed at the industrial scale on time. Regrettably, however, the global infrastructure associated with nuclear fuel R&D is contracting, and national regulations on nuclear research are consistently being reinforced. In such an environment, it would be nearly impossible for a single institution to acquire a comprehensive amount of experimental data covering all material properties and aspects of in-reactor performance required for timely fuel qualification. Therefore, cooperation with international organizations (IAEA, NEA) and advanced infrastructure would be essential for the successful implementation of advanced fuel technology. In a situation, where conducting an integral test using irradiated fuel rods or assemblies is challenging, a more practical alternative approach can be considered. Institutions with expertise perform unit tests using micro-machined samples to generate high-quality experimental data. A computational platform can then reliably predict the fuel performance of nuclear fuel by integrating unit data, constructing fuel behaviour models, and employing advanced computational science.

ACKNOWLEDGEMENT

This work was supported by a National Research Foundation of Korea (NRF) grant funded by the Korean government (grant number 2017M2A8A5015056).

REFERENCES

[1] KAERI.RE.KR, Korea Atomic Energy Research Institute 60 years history (June 2024), https://www.kaeri.re.kr/board?menud =MENU00433

[2] YANG, J. H., et al., Recycling process for sinter-active U3O8 powder, J. Nucl. Sci. Technol. **47** (2010) 538-541

[3] RHEE Y. W., et al., Fabrication of sintered annular fuel pellet for HANARO irradiation test, J. Nucl. Sci. Technol. **47** (2010) 345-350

[4] KIM, H.G., et al., Development status of accident tolerant fuel for light water reactors in Korea, Nucl. Eng. Technol. **48** (1) (2016) 1-15

[5] KIM D-J., et al., Fabrication of micro-cell UO2–Mo pellet with enhanced thermal conductivity, J. Nucl. Mater. **462** (2015) 289-295

[6] YANG, J. H., et al., UO2-UN composites with enhanced uranium density and thermal conductivity, J. Nucl. Mater. **465** (2015) 509-515

ID#23
CURRENT STATUS NUCLEAR FUEL PROGRAM IN INDONESIA

G. RAHMADI, G. K. SURYAMAN, P. VANKABO, H. GUFRON, A. ISNAINI, R. LANGENATI
National Research and Innovation Agency (BRIN),
Serpong, Indonesia

Abstract

This paper provides an overview of the current status of the Nuclear Fuel Program in Indonesia, with a focus on the NPP nuclear fuel initiative and research reactor fuel development. The investigation delves into the advancements, challenges, and key developments within the nuclear fuel programmes in Indonesia, encompassing both NPP and research reactor contexts. By examining the existing state of affairs, this study aims to contribute valuable insights into the trajectory and prospects of nuclear fuel endeavours in Indonesia, shedding light on the nation's efforts in the field of nuclear energy.

1. INTRODUCTION

Indonesia currently operates three research reactors; however, as of now, there is no operational power reactor in the country. Despite the absence of an NPP, Indonesia possesses fuel fabrication facilities for R&D purposes. The R&D activities related to fuel encompass both nuclear power reactor fuel and research reactor fuel [1] (see Figure 1). The R&D for nuclear power reactor fuel in Indonesia is conducted under the auspices of the Research Centre for Nuclear Material and Radioactive Waste Technology – National Research and Innovation Agency (BRIN) [1].

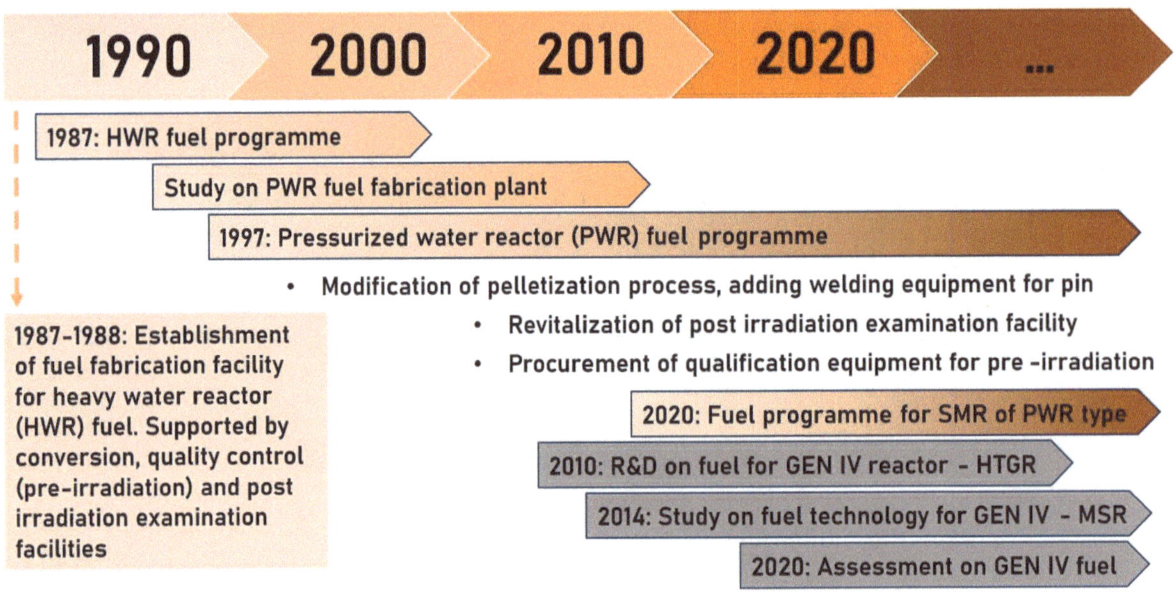

FIG. 1. History of NPP Fuel programmes.

The initiative for R&D on nuclear power reactor fuel in Indonesia commenced around 1987 with the HWR of Cirene. All equipment related to fuel production was installed in Serpong, Tangerang Selatan, within the same area as the G.A Siwabessy Research Reactor. In line with the evolution of reactor technology, Indonesia's interest shifted towards PWR technology in 1997, redirecting the focus of R&D from HWR to both PWR reactors and their corresponding fuels. In 2014, Indonesia initiated the Research and Development of Experimental Power Reactor (RDE) programme, targeting a 10 MW reactor based on HTGR technology. Consequently, fuel research shifted towards pebble-type HTGR fuel. However, research on PWR fuel continues unabated, reflecting its perceived potential for future development in Indonesia.

Indonesia is home to three research nuclear reactors: Kartini, TRIGA MARK, and Reaktor Serbaguna G.A. Siwabessy (RSG GAS). Nuclear fuel is a critical element in the operation of research reactors. Indonesia is capable of domestically producing fuel for the RSG GAS research reactor. Previous R&D efforts include U_3Si_2–Al with a density of 2.96 gU/cm^3, which is currently used in the reactor. Ongoing research aims to enhance its density to 4.8 and 5.2 gU/cm^3. Since 2018, Indonesia has been exploring the development of research reactor fuel based on UMo/Al and UZr/Al.

2. CURRENT STATUS NPP FUEL PROGRAMME IN INDONESIA

2.1 PWR Fuel Programmes

2.1.1 UO_2 Pellet

At present, research on the fabrication of nuclear fuel elements for power reactors primarily involves the use of sintered UO_2 pellet as the main nuclear fuel material. In addition to utilizing natural uranium, R&D on nuclear fuel fabrication at the Research Center for Nuclear Material and Radioactive Waste Technology is progressing by incorporating enriched uranium with enrichment levels ranging from 2% to 5%. The enrichment levels chosen for R&D align with the enrichment requirements for commercial nuclear reactor fuel, as seen in the United States, where enrichment levels typically range from 3–5% [2].

This research activity involves the use of UO_2 powder enriched at levels of 2%, 3%, 4%, and 5%. A portion of the powder is randomly selected from the UO_2 powder storage facility for observation of its characteristics, including the O/U ratio, bulk density, tap density, powder size distribution, and moisture content. Some of the powder is then supplemented with 0.4% by weight of Zinc Stearate and mixed using a blending tool for 30 minutes. The resulting mixture is compacted directly using a press machine. The raw pellets produced by the press machine undergo measurements and characterizations, including dimensions, weight, and density.

The measured raw pellets then undergo the sintering process under the following conditions: an initial heating rate of 250°C/hour in a N_2 atmosphere until reaching 700°C, after which the atmosphere is replaced with N_2+H_2 gas. The heating is continued at a rate of 250°C/hour until reaching a temperature of 1700°C, and the temperature is maintained for 3 hours. The cooling process involves a rate of 150 °C/hour until reaching 700 °C, where the atmosphere is changed to N_2 gas, followed by further cooling to room temperature. The sintered pellets produced undergo measurements and characterizations, including dimensions, weight, density, and O/U ratio.

The sintered pellets produced, as depicted in Table 1, for UO_2 with enrichments of 3% and 4% exhibit cracks or breakage, rendering the pellet dimensions unmeasurable. Meanwhile, for UO_2 with enrichments of 2% and 5%, intact sintered pellets were obtained without any signs of cracking or breakage. Measurements of dimensions and density calculations yielded a density of 10.02 g/cm^3 (91.36% Theoretical Density (TD)) for 2% enrichment and 8.35 g/cm^3 (76.07% TD) for 5% enrichment. The fuel requirements for PWR stipulate a sintered pellet density of more than 90% TD. All four types of produced sintered pellets present relatively low O/U ratios, nearly approaching 2. This indicates that sintering in an atmosphere of 2% H_2 and 98% N_2 is sufficient to reduce UO_2 with O/U ratios ranging from 2.34 to 2.48 to values between 2.0008 and 2.0051. For UO_2 fuel requirements, the O/U ratio has to be less than 2.015.

TABLE 1. UO_2 PELLETS AFTER CHARACTERIZATION [2]

	Enrichment 2%	Enrichment 3%	Enrichment 4%	Enrichment 5%
Total pellets	6	5	5	5
Pellet condition	No crack	crack	crack	No crack
Average height, mm	9.78	-	-	10.26
Average Diameter, mm	9.18	-	-	9.41
Theoretical Density,	91.36%	-	-	76.07%
O/U ratio	2.0012	2.0011	2.0008	2.0051

2.1.2 *PWR Fuel Pin programme*

An outline of the PWR fuel pin programme is illustrated in Table 2. The execution of this programme follows a structured approach, progressing from a dummy pin to a functional pin that contains enriched UO_2. The initial testing of the dummy pin spanned a decade, primarily attributed to challenges in procuring enriched uranium and Zircaloy-4, ensuring the readiness of Power Ramp Test Facility, and undertaking refurbishment work in Radiometallurgy Installation hot cells. Qualification has been conducted on specific parameters, with the intention to explore additional parameters as more infrastructure and equipment become accessible. The design for the preparation of the Stage 3 fuel pin prototype is currently in progress.

TABLE 2. PWR FUEL PIN PROGRAMME IN INDONESIA [1]

Stage	Prototype of PWR Fuel Pin	Programme Activities				Remarks
		Fabrication	Pre-irradiation	Irradiation	Post-Irradiation	
1	Dummy pin, without UO_2 pellet)	Prepared in 2013	Testing for selected parameters	Completed for 350 hours	Testing for selected parameters in 2016-2021	Feedback for design of fuel pin with enriched UO_2
2	Fuel pin with natural UO_2 pellet	Prepared in 2015	Testing for selected parameters	Completed > 1000 hours; Pending testing to reach 2000 hours	-	Awaiting readiness of Power Ramp Test Facility to resume irradiation testing
3	Fuel pin with enriched UO_2 at 2–5% U-235	At R&D and design stage (2020–2023) for preparation in 2024	-	-	-	Intensifying modelling to predict behaviour of spent fuel.
4	Improved design of prototype	Improved process parameters	Increased capacity for additional parameters	Varying irradiation parameters	Increased testing capacity for additional parameters	Option for benchmarking with partner

One crucial aspect of R&D is fuel fabrication. In this context, efforts have been undertaken to prepare fuel pin elements tailored for testing in PWR-type NPP reactors. The design of the fuel pin shape is adjusted to the testing facility, namely PRTF RSG-GAS. The primary material utilized in this experiment is enriched UO_2, and for structural components such as cladding and end caps, a specialized material known as zirconium alloy or Zircaloy is employed.

The cladding of the pin is prepared by cutting and shaping it from Zircaloy-4 casing material according to the design illustration. To achieve a good tip shape and the specified length, machining is performed using a lathe machine. The inner and outer parts of the casing are cleaned from dirt and grease with alcohol or wash benzene. Subsequently, ultrasonic cleaning with 25% alcohol is applied, and the final step involves measuring the casing dimensions.

The end cap is manufactured by turning a Zircaloy-4 rod. The Zircaloy-4 rod is shaped into upper and lower end caps according to the design illustration. Machining is conducted with a lathe machine to ensure uniform shape and size. To eliminate grease or other contaminants, cleaning is carried out similarly to the casing cleaning process. Following that, dimensions are measured for the lower cap, upper cap, and other components.

2.2 HTGR Fuel Programmes

2.2.1 Kernel Fabrication

According to the HTGR design adopted by the National Research and Innovation Agency for Nuclear Energy (ORTN BRIN), the fuel elements consist of pebbles with a diameter of 6 cm containing UO_2 kernels with a diameter of 0.5 cm. These kernels are coated with layers of pyrolytic carbon and silicon carbide, arranged in the following order from the innermost to the outermost layers: 1) a high-density pyrolytic carbon layer, 2) a silicon carbide layer, 3) a high-density pyrolytic carbon layer, and 4)a high-porosity pyrolytic carbon layer. These layered particles, known as TRISO particles, are uniformly distributed within a carbon matrix (natural graphite and electro-graphite), with the layered particle spaces located at the outermost layer. In the use of kernels for research and development of TRISO coating, consideration is given to the use of surrogate materials based on their ease of synthesis.

The use of surrogate UO_2 material for kernel purposes is intended for utilization in the research and development of TRISO coating processes. Zirconium, supplemented with doping materials to stabilize the zirconium dioxide phase at high temperatures, is employed in this research, considering that the TRISO coating process operates within the temperature range of 1250–1500°C. Additionally, this research aims to evaluate the fabrication process methodology, test the characteristics of the coating resulting from the coating process, and conduct heat tests on TRISO particles.

In the year 2023, the research on mastering HTGR fuel fabrication technology focuses on aspects such as feedstock preparation for zirconia kernel fabrication, including the degree of neutralization, stirring duration, and viscosity variations.

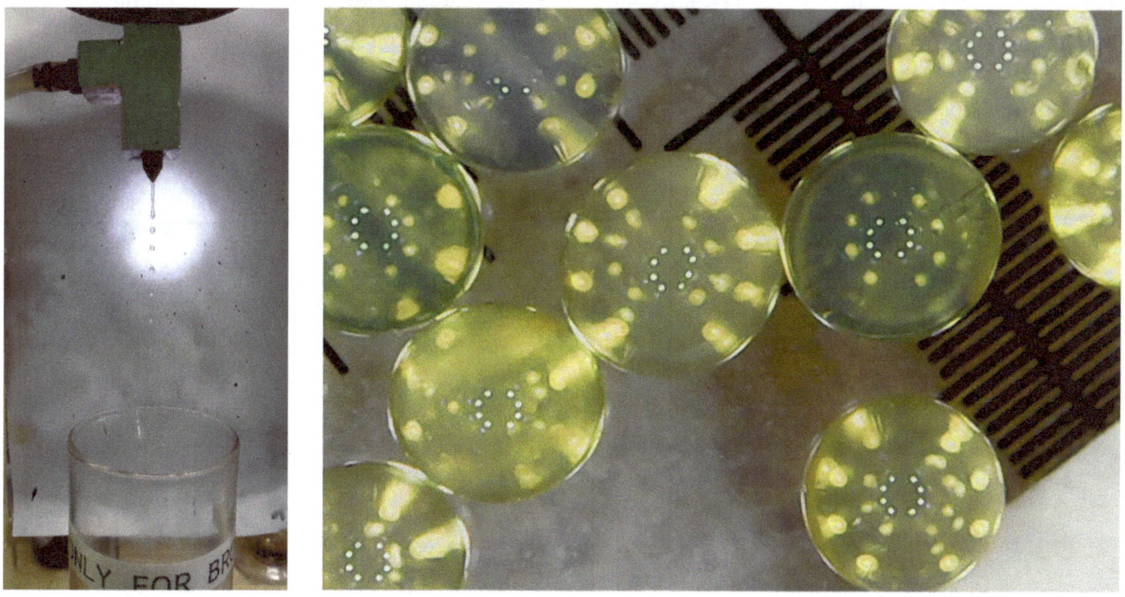

(a) (b)

FIG. 2. (a) Casting process, (b) wet microsphere after ageing process.

3. CURRENT STATUS RESEARCH REACTOR FUEL PROGRAMME IN INDONESIA

3.1 Research Reactor Fuel Programmes

3.1.1 UMo Fuel Fabrication Research

The UMo alloy material serves as a focal point in the ongoing research and development programme for advanced research reactor fuel in Indonesia, aligning with global efforts in this field. Experimental fabrication trials of mini-sized plate-type dispersion fuels containing UMo/Al, UMo/Al-Si, UMo-xM/Al, and UMo-xM/Al-Si (utilizing depleted uranium) have been conducted to identify the optimal process parameters. The aim is to explore the fabrication process parameters for subsequent application in the production of plate-type dispersion fuels with 19.75% ^{235}U uranium enrichment for irradiation testing.

Uranium-based alloys, particularly uranium–molybdenum (U-Mo), play a crucial role in nuclear engineering as potential nuclear fuel for research reactors. UMo alloys with Mo content ranging from 7 to 10% by weight exhibit promising prospects for utilization as LEU dispersion nuclear fuel. The UMo alloy density surpasses 16.0 g/cm^3 (depending on Mo content) and possesses corrosion-resistant properties. However, due to its ductile nature, UMo alloy encounters challenges in mechanical powder production processes such as grinding mills or ball mills. This density is higher compared to uranium/uranium oxide alloys traditionally used as research reactor fuels, such as U_3Si_2, U_3O_8, and UAlx, with densities of 12.2, 8.4, and 6.7 g/cm^3, respectively. Employing U–Mo-based dispersion fuel enhances the uranium density within the core (meat) of fuel plate elements to exceed 8 gU/cm^3, meeting the requirements for research reactors employing LEU.

The metal molybdenum (Mo) exhibits a density of 10.2 g/cm^3, a low neutron absorption cross-section (2.65 barn), a melting point of 2620°C, and high solid solubility with γ-U. Under slow cooling or for alloys containing less than 7% Mo, the equilibrium state of U(Mo) alloys at temperatures under 560°C characterized by a combination of αU and $\gamma'(U_2Mo)$ phases. At higher temperatures, however, the solid solution of γ-U/Mo is formed. The addition of Mo in the range of 5–20% Mo atoms serves to stabilize the γ-U phase while maintaining uranium density and enhancing the fuel's melting point. Based on these properties, U–Mo-based alloys are promoted as candidates for dispersion fuel, as they theoretically can be fabricated with very high uranium density and post-irradiation fuel processing can be easily carried out.

The U–7Mo–xM alloy is produced through a melting process using an electric arc furnace with argon gas as the medium. Melting is carried out with a current of 150 A, and each alloy undergoes five repetitions of the melting process. The resulting alloy ingots are sufficiently homogeneous, ductile, and visually free from observable uranium oxide on the surface. The ductile nature of the ingots prevents them from being directly transformed into powder through mechanical means. Therefore, an alternative method, specifically the hydriding-dehydriding-grinding mill technique, is employed for powder production.

The U–7Mo–xM alloy, obtained through the dehydriding process, exhibits brittleness and high reactivity towards oxygen, leading to the formation of uranium oxide. Consequently, powder production involves the use of a grinding mill, and an enrichment process is employed to separate particle grains using standard ASTM sieves within a glove box filled with argon gas.

The U–7Mo–xM powder (see Figure 3) subjected to the sieving process is then analysed for U content, impurities, and density. The U content and density analysis results for the U–7Mo–xM powder are presented in Table 2. The analysis indicates that with an increase in the concentration of the third element in the alloy, both the U content and density decrease. The particle size distribution of the powder can be controlled by adjusting the grinding mill process time, resulting in a composition of -90 +38 µm = 75–85% and -38 µm = 15–25%, meeting the specifications for RSG-GAS fuel material.

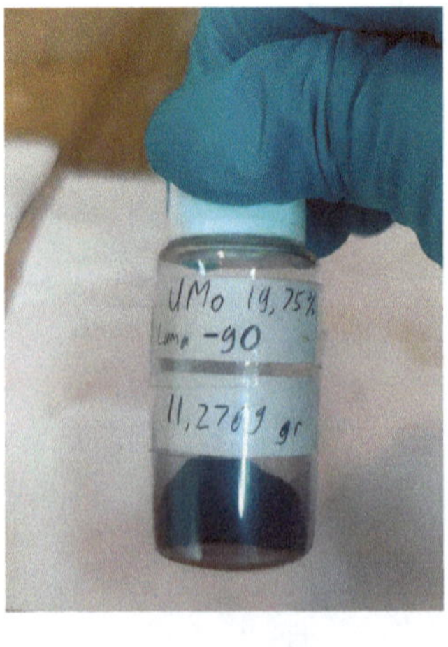

FIG. 3. U–7Mo Powder.

The frame and cover plates are manufactured through machining using the $AlMg_2$ alloy. The fabrication of the frame and cover involves the rolling of $AlMg_2$ sheet plates followed by machining. The $AlMg_2$ sheet rolling for the frame is conducted until a thickness of 3.15 mm is achieved, while for the cover, it is rolled to a thickness of 2.7 mm.

Subsequently, the $AlMg_2$ sheet with a thickness of 3.15 mm is machined to create the frame, sized [180 mm × 140 mm × 3.15 mm], with a centre hole measuring 25 mm × 15 mm × 3.15 mm. On the other hand, the plate with a thickness of 2.7 mm is machined to form the cover, sized [180 mm × 140 mm × 2.7 mm] [3].

The U–7Mo–xM powder blend and matrix powder are mixed in a weight ratio to achieve a uranium density of 7 gU/cm^3. The mixture is homogenized and shaped into U–7Mo–xM/Al fuel element cores through pressing at pressures ranging from 10 to 20 bar. The resulting pressed fuel element cores are rectangular with dimensions conforming to the design specifications, measuring 25 mm in length and 15 mm in width. However, observations reveal variations in thickness, with some measurements outside the specified range (3.15 ± 0.05 mm). Visual inspection indicates satisfactory pressing results for the U–7Mo–xM/Al fuel element cores, showing homogeneity and an absence of cracks.

From the measurements, it is evident that the length and width adhere to the design specifications, while thickness exhibits variations beyond the 3.15 ± 0.05 mm requirement in some fuel element cores. The thickness deviations outside the designated range may be attributed to an uneven distribution of the matrix powder. A crucial consideration in practice is that if the uranium content within the fuel element cores meets the specified requirements, adjustments to the Al matrix can be made to achieve the desired fuel element thickness.

4. CONCLUSIONS

The investigation highlighted the nation's efforts to address the complexities associated with fuel fabrication, and safety considerations. Insights into the progress, challenges, and prospects of NPP fuel programmes in Indonesia were explored, contributing valuable information to the broader discourse on nuclear energy development.

In parallel, the research reactor fuel development in Indonesia was scrutinized, shedding light on the advancements and challenges encountered in this specific domain. The exploration of materials, manufacturing techniques, and QC measures underscored the nation's commitment to advancing research reactor fuel technologies.

The analysis presented in this paper emphasizes the importance of continued research, development, and collaboration in Indonesia's nuclear fuel programmes to meet the growing energy demands while ensuring safety, security, and sustainability. As Indonesia navigates its nuclear energy journey, lessons learned from both NPP, and research reactor fuel initiatives will play a crucial role in shaping the trajectory of the nation's nuclear fuel landscape. The findings presented herein aim to contribute to the broader understanding of Indonesia's nuclear energy ambitions and foster informed decision-making for the future.

REFERENCES

[1] JOHARI, J. M. C., et al., Approach toward strengthening strategy for nuclear fuel development in Indonesia (Proc. InAIP Conf., 2022), Vol. 2501, No. 1, AIP Publishing (2022).

[2] YULIANTO, T., Pembuatan pelet Sinter UO2 diperkaya dan penyiapan komponen perakitan pin uji elemen bakar PWR, in Prosiding Seminar Hasil-hasil Penelitian EBN (2017) 76.

[3] SUPARDJO, S., et al., Pengembangan teknologi bahan bakar reaktor riset berbasis U-Mo/Al [Fabrikasi dan Pengujian Pelat Elemen Bakar U-Mo-xM (M= Ti, Zr, Si)], in Prosiding Seminar Hasil-hasil Penelitian EBN Tahun 2017 (2018) 116.

ID#24

CURRENT STATUS NUCLEAR FUEL PROGRAMME IN FRANCE MOX

G. KERBOUL, J-M. MARIN
Orano,
France

Abstract

The paper provides an overview of MOX fuel fabrication at Orano and gives some comparison elements between MOX fuel and UO_2 fuel. The R&D programme on the multi-recycling of used fuels is also discussed.

1. INTRODUCTION - MOX FABRICATION AT ORANO

Orano is a recognized international leading operator in the field of nuclear materials and offers its customers high performing products and services throughout the entire fuel cycle, i.e. mining, conversion, enrichment, recycling, logistics, engineering and decommissioning.

By recycling, Orano recovers useful materials (U and Pu) from spent fuel and reuses them in new fuels such as MOX, and thus in nuclear reactors. Used fuel reprocessing operations are performed at Orano La Hague facility and MOX fuel fabrication at Orano Melox facility.

MOX fuel is composed of an average of 8.5% plutonium, mixed with depleted uranium. MOX fuel is then not only a factor of non-proliferation because the plutonium is not put in the final disposal as is the case once-through cycle, but also a way to reuse a byproduct of our enrichment operations, i.e. depleted uranium.

FIG. 1. Aerial view of Melox.

Since 1972, 44 reactors have been loaded worldwide with MOX fuels. In France every year 10 tonnes of plutonium are recycled, and it represents 10% of the nuclear electricity produced in France.

Melox (Figure 1), the MOX fuel fabrication plant (whose manufacturing process is based on the MIMAS process - MIcronisation MASterblend), started the industrial MOX fuel production for French PWR reactors in 1995.

In 1997, the plant reached its initial licensed capacity of 100 tHM/year. After a qualification period, it was possible to manufacture MOX fuels for foreign customers (Japan and Germany). In order to support this international deployment, investments to increase capacity have been made, and the operating decree of the Melox plant has been amended to authorize an increase in production to 145 tHM/year in 2003 and 195 tHM/year in 2007.

Since 1995, more than 3000 tonnes of MOX have been produced until the end of 2023, and in 2011, more than 1,000,000 fuel rods have been fabricated and certified.

Melox delivers its MOX fuels to 22 over the 58 French NPPs. They all are 900 MW reactors, loaded with 30% MOX in their core power. Orano has also delivered MOX fuels to Belgium, Netherlands, Germany, Switzerland, Japan and to the USA, where one reactor was loaded with four FAs a few years back.

2. MOX DESIGN COMPARED TO UO_2

UO_2 and MOX FAs have an identical fuel mechanical design: same length, same rod diameter, same cladding thickness, same material and same pellet cladding gap.

One difference concerns the fuel pellets:

- In UO_2 fuel the uranium fissile material, enriched ^{235}U, is homogeneously present in the fuel pellet;
- In MOX fuel the fissile material, plutonium, is blended to a uranium carrier material (depleted uranium) (see Figure 2).

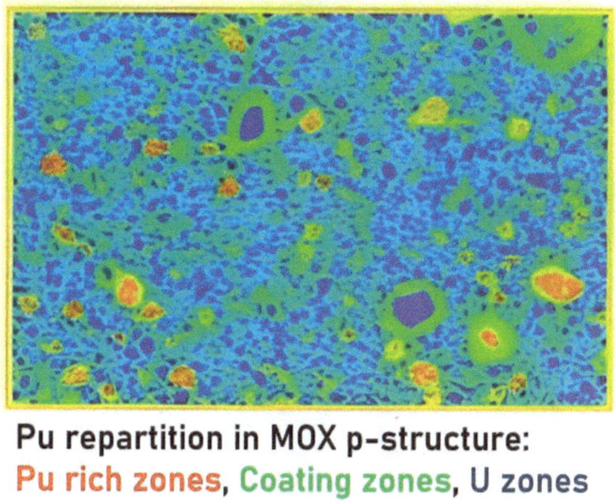

FIG. 2. Pu repartition in MOX p-structure.

The MOX fuel specificities result in different behaviour in reactor compared to UO_2 fuel. During irradiation, for example, the fuel rod growth, the cladding deformation and corrosion are slightly different but not to a significant extent requiring a design change for the reactor.

As the higher fission cross section of MOX fuel may cause a power peak at UO_2/MOX interfaces, the design of the MOX FAs was adapted by creating a 3 zones design (high, medium and low Pu content) in order to flatten the power distribution at the interfaces (see Figure 3).

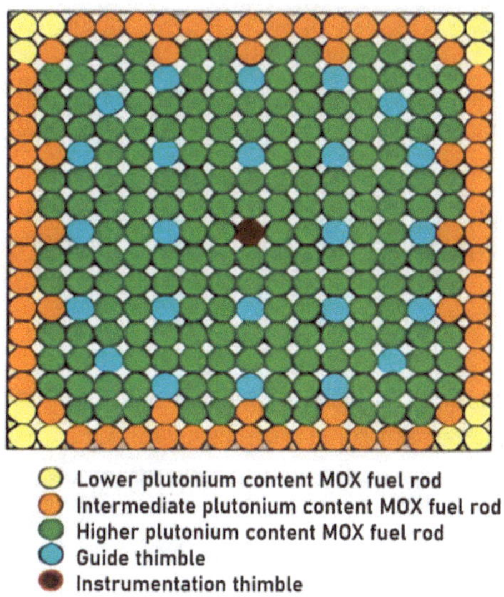

FIG. 3. Zones with different Pu content used in MOX fuel.

The other difference between UO_2 and MOX fuel concerns the plenum volume. As the fission gas release, which can contribute to internal pressure increase, is higher with MOX fuel, the rod plenum volume has been optimized.

3. MOX PERFORMANCES

Mixed oxide fuels have been used in 44 reactors since 1972 with a perfect safety track record for more than 40 years. Mixed oxide fuels are adapted to utilities and power plants' requirements and their evolution. For example, MOX fuel is capable of either base load or load follow. In terms of burn up, MOX fuel is following the trends of uranium fuel. The MOX performances are the same or equivalent to uranium.

Moreover, as shown by Figure 4, MOX fuel has been licenced by 7 national safety authorities.

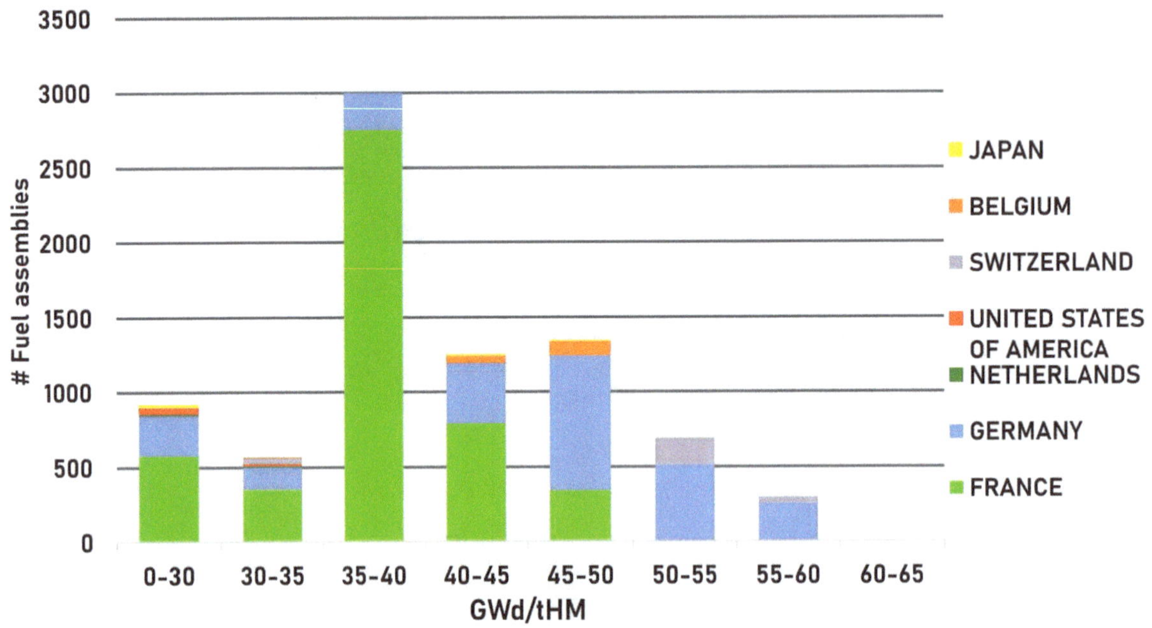

FIG. 4. Discharged burn up of MOX FAs in different countries.

4. FABRICATION PROCESS

MOX fuel pellets are produced using the MIMAS process, meaning MIcronization of a MASter blend. This is a two-step process to produce a high plutonium content master blend that will be diluted and mixed with a uranium dioxide powder in order to manufacture (U, Pu)O_2 fuel pellets, in the range of 2% to 12% of Plutonium content. After blending, the process is very similar to a uranium fuel fabrication process. The MOX pellets are cladding in the same type of rods and the same type of FAs as the ones used for UO_2 fuel.

5. SAFETY

In terms of safety, for prevention and control of nuclear material, especially dispersion and exposure to operator workers, a containment system with three barriers under negative air pressure is used: first the glove boxes under nitrogen atmosphere, the room and around a specific building which is leak tight and in negative pressure compared to the external.

The major risk prevention has been integrated into the design of the facility (fire, criticality, seismic and thermal risks).

6. MOX FUEL PERSPECTIVES

The French multiannual Energy Program will be deployed as follows:

— Current mono-recycling strategy at least until 2040;
— Preparation of MOX fuelling of part of French 1300 MW reactors to accommodate planned shutdowns of some 900 MW reactors and ensure the robustness of the system.

To support the closure of the nuclear fuel cycle, a R&D programme is currently focusing on the multi-recycling of used fuels in PWR in the mid-term, maintaining the perspective of potential industrial deployment of a fleet of fast neutron reactors in the long term.

Reprocessing irradiated MOX fuels to recover the second generation of plutonium for multi-recycled MOX fuel manufacturing will allow to stabilise used fuel inventory and control of plutonium inventory.

However, the challenge is that the isotopic composition of multi-recycled plutonium is evolving towards a decrease in fissile material.

At least, for the long term, the strategic industrial and R&D roadmap leads to the fuel cycle closure and the development of GEN IV reactors, fast neutron reactors using solid fuels or molten salt reactors, to balance the inventory of the used fuel and also of the separated plutonium.

7. CONCLUSION

Plutonium recycling through fabrication at the Orano Melox plant and irradiation of MOX fuel in the LWRs have demonstrated the reliability of this process,

The operational performance of MOX fuel is as good as uranium fuel.

To stabilize the number of used MOX FAs and prepare to close the cycle with GENIV reactors, multi recycling strategies are studied.

The flexibility of Melox plant helps planning for the future: continuous adaptations are developed and tested to prepare implementation of future concepts.

The decrease in fissile quality is followed by a higher dose level. Specific developments are also underway to better protect workers against irradiation.

ID#25
SMR FUEL DEVELOPMENT AND FABRICATION – INB OVERVIEW

P. O. SOUZA
Indústrias Nucleares do Brasil (INB),
Rio de Janeiro, Brazil

Abstract

The paper provides a basic comparison of fuels for SMRs and fuels already manufactured at the Brazilian company Indústrias Nucleares do Brasil. It shows the company's capabilities and impacts expected in the production processes.

1. INTRODUCTION

The INB – Indústrias Nucleares do Brasil is a state owned (public company) of Brazil linked to the Ministry of Mines and Energy. So, we exercise, on behalf of the Union, the monopoly on the production and sale of nuclear materials. We also operate in the execution of fuel engineering services and in the production of fuel element components.

The Brazilian Nuclear Regulatory Body responsible for licensing and supervising the INB's activities is the National Nuclear Energy Commission (CNEN).

INB's operations on nuclear fuel cycle include mining, milling, enrichment, reconversion, manufacturing of pellets and the FA. With units in Resende (RJ), Buena (RJ), Caetité (BA), Caldas (MG), and São Paulo, INB has its headquarters in the city of Rio de Janeiro and an office in Fortaleza (CE), the base of the Santa Quitéria Project.

Currently, we supply the demand for nuclear fuel for the Brazilian NPPs – Angra 1 and Angra 2, moreover, FAs for the first core of Angra 3 (under construction).

The Angra 1 plant is a PWR with 640 MWe (Westinghouse technology) – and uses 121 FAs of 16NGF design. The Angra 2 plant is a PWR 1350 MWe (Siemens technology) – and uses 193 FAs of HTP-I design. The Angra 3 plant will be similar to Angra 2 NPP with 1405 MWe.

Given the growing demand for energy and targets established for Brazil, from 2 GW in 2023 to 8–10 GW by 2050 of nuclear power generation, according to an official report of the Brazilian Energy Research Office (EPE in its Portuguese acronym). Combined with the 2050 Net Zero scenario, it can be stated that nuclear power plays a fundamental role in Brazil outlook of establishing a system without increasing carbonization.

In order to meet this future internal demand and also make part of the international fuel cycle market, the planning and challenges for nuclear fuel production in Brazil involve investments, increasing production capacity, and licensing.

2. SMALL MODULAR REACTOR

Talking about worldwide, today we witness a wave of innovation in SMRs. Driven mainly by the desire of the global community to reduce emissions of carbon and the need to assure energy security for all countries and people.

Although several challenges have to be addressed, such as regulatory frameworks, public acceptance, and ensuring the secure handling of nuclear materials, the SMRs present a promising solution for sustainable energy generation and INB has the potential to play a significant role in the future of SMR fuel fabrication landscape.

There are several SMRs under development with a variety of designs, from water cooled, gas cooled, liquid metal, and molten salt to microreactors, land based or marine based and LEU to HALEU. As, recently listed in Ref. [1] more than 80 SMRs, including microreactors are in some stage of development. The Nuclear Energy Agency reported in its dashboard of 2023 [2] different SMR designs with fuel enrichment from 5% (LEU) to 20% (HALEU) at ranges of thermal power mostly around 150–600 MW_{th}.

2.1 Brazilian Scenario

A study performed by the Idaho National Laboratory, in collaboration with the United States Department of Energy and the Brazilian Office EPE [3] explored the suitability of power reactor designs for the Brazilian market and selected four design categories that are under consideration by EPE, including a thorium-fuelled reactor design, since Brazil has an interest in utilizing domestic thorium resources. The other three types of reactors include LWR-type SMR, high-temperature gas-cooled SMR, and Microreactor. In addition, an important factor to be highlighted is the recent inclusion of the advancement of SMRs as one of the priorities in the industrial policies of the current Brazilian government.

Despite the absence of complete understanding regarding certain technical aspects for determining an optimal design for future Brazilian reactors, one research was done here considering the LWR-type (land based) SMR designs regarding the fuel array, length, enrichment, and development status, with the aim of evaluate types of SMR fuels that could be manufactured at INB in the near future. This research led to the selection of the following types of SMR presented in Table 1.

TABLE 1. TYPES OF SMR

Acronym	Vendor	Reactor Type
VOYGR™	NuScale Power, Inc.	Integral PWR
NUWARD™	CEA, EDF, Naval Group and TechnicAtome	Integral PWR
SMR-160	Holtec International	PWR

2.2 Main characteristics of SMRs

The data available at the IAEA Advanced Reactors Information System (ARIS) database together with Reference [1], was used to perform a comparison of the main technical parameters for the three SMRs under evaluation with the current nuclear fuel produced by INB (Table 2).

TABLE 2. MAJOR TECHNICAL PARAMETERS

	NUSCALE (VOYGR™)[a]	HOLTEC SMR-160	NUWARD™	ANGRA 2	ANGRA 1
Thermal Power (MWth)	250	525	2 × 540	3765	1876
Electrical Power (MWe, gross)	77[b]	174	2 × 170	1350	640
Primary circulation	Natural		Forced		
System pressure (bar absolute)	138	155	150	158	155
Core Inlet Temperature (°C)	249	243	280	290	290
Number of FAs in the core	37	57	2 × 76	193	121
Fuel Cladding Material	M5™	M5™	Zr alloy	M5™	ZIRLO[c]

TABLE 2. MAJOR TECHNICAL PARAMETERS (CONT.)

FA array	Standard LWR fuel in a 17 × 17 square			16 × 16 square	
Fuel Design	Modified HTP2™	GAIA[d]	-	HTP-I	NGF
Fissile column length (m)	2.0 (shortened core height)	standard PWR length	2.2 (shortened core height)	3.9	3.66
Solid Burnable absorber	Gd_2O_3	-	Gd_2O_3	Gd_2O_3	Gd_2O_3
Refuelling cycle (months)	18	24	24	14	14
Reactivity control	Control Rods, soluble Boron		Control Rods, Boron-free	Control Rods, soluble Boron	
Fuel	UO_2 ≤4.95%		UO_2 <5%	UO_2 ≤4.95%	

[a] NuScale Power and VOYAGR are trademarks or registered trademarks of NuScale Power, LLC.
[b] per module
[c] ZIRLO is trademark or registered trademark of Westinghouse Electric Company LLC, its Affiliates and/or its Subsidiaries in the USA and may be registered in other countries throughout the world.
[d] GAIA, HTP-I, M5 are trademarks or registered trademarks of Framatome or its affiliates, in the USA or other countries.

2.3 Manufacturing Possibilities at INB

The comparison done showed that the fuels for the SMRs of Table 2 have enrichments and a technology platform very similar to the NGF and HTP fuels currently manufactured by INB, with some modifications regarding the fuel rod length, FA array and components, such as nozzles and grids. So, besides the challenges of increasing production it is expected only a few adaptations to the technologies for manufacturing SMR fuels at INB factory, as follows:

- No/very low impact: Reconversion, Pelletizing, Enrichment;
- Slight impact: Skeleton Welding, Fuel Rod Assembling, Assembling Storage;
- Significant impact: FA Shipment, FA Storage.

3. CONCLUSIONS

So, it is concluded that the FA factory of Brazil has the potential to meet the demand of future SMR fuel fabrication, for all mentioned and other SMR technologies. Besides the Brazilian scenario of increasing the NPPs, the Latin America SMR market can also benefit from the proven capabilities of INB company.

REFERENCES

[1] INTERNATIONAL ATOMIC ENERGY AGENCY, Advances in Small Modular Reactor Technology Developments, A Supplement to: IAEA Advanced Reactors Information System (ARIS), Vienna (2022).

[2] NUCLEAR ENERGY AGENCY, The NEA Small Modular Reactor Dashboard, Boulogne-Billancourt, France (2023)

[3] UNITED STATES-BRAZIL JOINT STUDY: A Preliminary Assessment of Opportunities and Challenges for Small Modular Reactors in Brazil, Idaho Falls, Idaho (2023)

APPENDIX

MEETING PROGRAMME

2021 TECHNICAL MEETING SESSIONS

TECHNICAL SESSION I: Powder and Pelletizing

Presentation ID	Title	Presenter	Country	Org
ID#8	Fabrication and property evaluation of microcell and microplate UO_2 pellets	Jae Ho YANG	Korea, Republic of	KAERI
ID#9	Development and methodology or the characterization of the mechanical integrity of PHWR fuel pellets	Juan Pablo MEDINA	Argentina	CNEA
ID#4	MOX fuel process: Powders and pelletizing fabrication experience	Jean-Michel MARIN	France	Orano

TECHNICAL SESSION II: Fuel Rod Manufacturing and Inspection

Presentation ID	Title	Presenter	Country	Org
ID#24	Orano's MOX fuel experience and perspectives	Jean-Michel MARIN	France	Orano
ID#17	applications of machine vision technology in CANDU nuclear fuel production line	Feng HUO	China	CNNF
ID#16	Improvement of Homemade Coater	Diyang FAN	China	CNNF

TECHNICAL SESSION III: Fuel Assembly and Appendage Welding

Presentation ID	Title	Presenter	Country	Org
ID#12	Replacement of beryllium as a filler material in brazing of CANDU fuel	Francisco DIMAYUGA	Canada	CNL
ID#11	Alternatives to replace the use of beryllium for appendages attachment on CANDU fuel elements	Alejandro Anibal BUSSOLINI	Argentina	CNEA

TECHNICAL SESSION IV: Fuel Pellet Examinations

Presentation ID	Title	Presenter	Country	Org
ID#20	Novel compact high speed passive gamma ray assay equipment for uranium enrichment measurement of fuel rods	Yongli ZHU	China	CNNFC
ID#10	The UO_2 pellet Ceramography Evaluation and Engineering Ceramic Microstructure Analysis	Abdollah RIAHI	Iran, Islamic Republic of	AEOI

TECHNICAL SESSION V: Building Fuel Technology in Nuclear Embarking Countries

Presentation ID	Title	Presenter	Country	Org
ID#23	Current status of nuclear fuel programme in Indonesia	Gagad RAHMADI	Indonesia	BRIN
ID#27	Nuclear programme in Jordan	Amani ALMAHMOUD	Jordan	Energy and Minerals Regulatory Commission

2023 TECHNICAL MEETING SESSIONS

TECHNICAL SESSION I Powder and reconversion (Chair: J.H. Yang, KAERI, Republic of Korea)

Presentation ID	Title	Presenter	Country	Org
ID#3	Uranium oxide powder studies for nuclear fuel fabrication	Anne-Charlotte Robisson	France	CEA
ID#1	Development of ammonia uranate hydrate (AUH) wet reconversion process for the production of nuclear grade UO_2 powder from uranyl nitrate hexahydrate solution	Byungkuk Lee	Korea, Republic of	KEPCO NF
ID#2	Modelling of the dry conversion process	Jeremy Bischoff	France	Framatome

TECHNICAL SESSION II Powder Sintering, pelletizing, advanced fuel pellet fabrication (Chair: A-C. Robisson, CEA, France)

Paper ID	Title	Presenter	Country	Org
ID#6	Evaluation of the correlation between green and sintered density in UO_2 pellets	Joao Paulo Rodrigues Carnaval	Brazil	INB
ID#5	Modelling of the fuel sintering process	Gaëlle RAVEU	France	Framatome
ID#8	Development status of high thermal conductivity ATF pellet and burnable absorber fuel	Dong-Joo Kim	Korea, Republic of	KAERI
ID#7	Advances in fabrication and characterization of gadolinium pellets for SMR fuel in Argentina	Maria F. Parrado	Argentina	CNEA

TECHNICAL SESSION III: Cladding, fuel rod and assembly components (Chair: I. Dimayuga, CNL, Canada)

Paper ID	Title	Presenter	Country	Org
ID#14	Development status of ATF and 3D printing technology for nuclear	Hyun Gil Kim	Korea, Republic of	KAERI
ID#12	Developments in fuel fabrication at CNL	Ike Dimayuga	Canada	CNL
ID#13	Industrial development of N36 zirconium alloy cladding tubes	Chunrong Xu	China	NPIC

TECHNICAL SESSION IV: Advancements in fabrication facilities and inspection (Chair: J. Bischoff, Framatome, France)

Paper ID	Title	Presenter	Country	Org
ID#15	Development trends and recent achievements in fuel fabrication technology at TVEL facilities	Alexander Radostin	Russian Federation	TVEL
ID#18	New technologies and innovations in quality inspection equipment at ENUSA's factory	David Verdejo Garrido	Spain	ENUSA
ID#19	Double line passive scanner for uranium and gadolinium fuel rods inspection at ENUSA´s factory	Javier Lafuente Sevillano	Spain	ENUSA

TECHNICAL SESSION V: Advancements in fabrication facilities and inspection (Chair: K.S. Sim, IAEA)

Paper ID	Title	Presenter	Country	Org
ID#26	Implementation of a new type nuclear fuel at the Armenian NPP	Arakel Arakelyan	Armenia	Haykakan Atomayin Metsamor
ID#27	SMR fuel development and fabrication – INB Overview	Patricia Oliveira de Souza	Spain	INB

REFERENCES

[1] INTERNATIONAL ATOMIC ENERGY AGENCY, Review of Fuel Failures in Water Cooled Reactors (2006-2015), IAEA Nuclear Energy Series No. NF-T-2.5, IAEA, Vienna (2019).

[2] INTERNATIONAL ORGANIZATION FOR STANDARDIZATION, Nuclear fuel technology - Guidelines for ceramographic preparation of UO2 sintered pellets for microstructure examination, ISO 16793:2018, ISO, Geneva (2018).

[3] AMERICAN SOCIETY FOR TESTING MATERIALS, Standard Practice for Ceramographic Preparation of UO2 and Mixed Oxide (U,Pu)O2 Pellets for Microstructural Analysis, ASTM C1868-18, ASTM International, West Conshohocken, PA (2018).

[4] INTERNATIONAL ATOMIC ENERGY AGENCY, Quality and Reliability Aspects in Nuclear Power Reactor Fuel Engineering - Guidance and Best Practices to Improve Nuclear Fuel Reliability and Performance in Water Cooled Reactors, IAEA Nuclear Energy Series No. NF-G-2.1 (Rev. 1), IAEA, Vienna (2024).

[5] WORLD NUCLEAR NEWS, TVEL puts case at conference for TVS-K performance, (August 2024), https://www.world-nuclear-news.org/Articles/TVEL-tells-conference-of-TVSK-fuel-plans.

[6] WORLD NUCLEAR NEWS, EU funding Framatome VVER-440 fuel development, (August 2024), https://www.world-nuclear-news.org/Articles/EU-funding-Framatome%C2%A0VVER-440-fuel-development.

[7] WESTINGHOUSENUCLEAR.COM, Westinghouse Creates and Installs Industry's First 3D-Printed Fuel Debris Filter for Nuclear Power Plants, (August 2024), https://info.westinghousenuclear.com/news/westinghouse-creates-and-installs-industrys-first-3d-printed-fuel-debris-filter-for-nuclear-power-plants.

[8] FRAMATOME.COM, Framatome installs first 3D-printed stainless steel fuel component at Forsmark Nuclear Power Plant, (August 2024), https://www.framatome.com/medias/framatome-installs-first-3d-printed-stainless-steel-fuel-component-at-forsmarknuclear-power-plant/.

[9] WESTINGHOUSENUCLEAR.COM, Westinghouse Produces 1,000th Additive Manufacturing Component for VVER-440 Fuel, (August 2024), https://info.westinghousenuclear.com/news/westinghouse-produces-1000-additive-manufacturing-component-for-vver-440-fuel.

LIST OF ABBREVIATIONS

ADS	Aluminium Distearate
AIP	Arc Ion Plating
ADU	ammonium diuranate
AEOI	Atomic Energy Organization of Iran
AI	Artificial Intelligence
AM	Additive Manufacturing
API	Automatic Pellet Inspection
ASTM	American Society for Testing and Materials
ATF	Accident Tolerant Fuels, Advanced Technology Fuels
AUC	ammonium uranyl carbonate
AUH	ammonium uranate hydrate
BET	Brunauer–Emmett–Teller
BRIN	National Research and Innovation Agency of Indonesia
BWR	boiling water reactor
CANDU	CANadian deuterium-uranium reactor
CEA	Commissariat à l'Énergie Atomique et aux Énergies Alternatives (France)
CNEA	Comisión Nacional de Energía Atómica (Argentina)
CNL	Canada National Laboratories
CNNFC	China North Nuclear Fuel Co. Ltd.
DC	Dry Conversion
DED	Direct Energy Deposition
DfAM	Design for AM
DLPS	Double Line Passive Scanner
DME	demountable fuel elements
DNN	Deep Neural Networks
FA	Fuel Assembly
GD	Green Density
GEN IV	Generation IV
HALEU	High assay low enriched uranium
HTGR	High Temperature Gas Reactor
HWR	Heavy water reactor
IAEA	International Atomic Energy Agency
IEC	Fuel Engineering Department
INB	Indústrias Nucleares do Brasil

ISO	International Organization for Standardization
KAERI	Korea Atomic Energy Research Institute
KEPCO NF	KEPCO Nuclear Fuel
LEU	Low Enriched Uranium
LEU+	Low Enriched Uranium (from 5 to 10%)
LOCA	Loss of Coolant Accident
LWR	Light Water Reactor
MIMAS	MIcronization of a MASter blend
MOX	Mixed Oxide Fuel
NPIC	Nuclear Power Institute of China
NPP	Nuclear power plant
ODS	Oxide dispersion strengthened
PBF	Powder Bed Fusion
PHWR	Pressurized Heavy Water Reactor
PLC	Programmable logic controller
ppb	parts per billion
ppm	parts per million
PWR	Pressurized Water Reactor
R&D	Research and Development
SD	Sintered Density
SEM	Scanning electron microscopy
SEU	Slightly Enriched Uranium
SMR	Small Modular Reactor
TD	Theoretical Density
TRISO	Tri-structural Isotropic
QA	Quality assurance
QC	Quality control
WWER	Water Water Energetic Reactor (or water cooled, water moderated energy reactor)(Russian PWR); also called VVER

LIST OF PARTICIPANTS

LIST OF PARTICIPANTS FOR 2021 TECHNICAL MEETING

ARGENTINA

Alvarez L.	Comisión Nacional de Energía Atómica (CNEA)
Lemos L. S.	Comisión Nacional de Energía Atómica (CNEA)
Medina J.	Comisión Nacional de Energía Atómica (CNEA)
Bussolini A. A.	Comisión Nacional de Energía Atómica (CNEA)

BANGLADESH

Motalab M. A.	Bangladesh Atomic Energy Commission (BAEC)

BELARUS

Sikorin S.	The State Scientific Institution "The Joint Institute for Power and Nuclear Research-Sosny"
Tretiakevich S.	Department of Nuclear and Radiation Safety of the Ministry for Emergency Situations of the Republic of Belarus (Gosatomnadzor)

BRAZIL

Abe A.	Instituto de Pesquisas Energeticas e Nucleares
Costa de Oliveira G.	Nuclear Industrial Center of Aramar
Gabriel Lopes J. M.	Nuclear Industrial Center of Aramar
Lazzari Garcia R. H.	Instituto de Pesquisas Energeticas e Nucleares

CANADA

Dimayuga F.	Canadian Nuclear Laboratories (CNL)

CHINA

Fan D.	China North Nuclear Fuel Co. (CNNFC)
Huo F.	China North Nuclear Fuel Co. (CNNFC)
Zhu Y.	China North Nuclear Fuel Co. (CNNFC)

EGYPT

Moustafa A.	Nuclear Power Plant Authority

FRANCE

Bargues S.	Orano China
Hatron P.	Orano China

Marin J-M.	Orano

HUNGARY

Lente B.	Hungarian Atomic Energy Authority
Nagy M.	Hungarian Atomic Energy Authority

INDIA

Deo A. K.	Atomic Energy Regulatory Board

INDONESIA

Johari J. M. C.	National Research and Innovation Agency (BRIN)
Rahmadi G.	National Research and Innovation Agency (BRIN)

IRAN, ISLAMIC REPUBLIC OF

Nazem H.	Atomic Energy Organization of Iran (AEOI)
Riahi A.	Atomic Energy Organization of Iran (AEOI)

JORDAN

Almahmoud A.	Energy and Minerals Regulatory Commission (EMRC)

KOREA, REPUBLIC OF

Yang J-H.	Korea Atomic Energy Research Institute (KAERI)

PAKISTAN

Shah F. S.	Pakistan Atomic Energy Commission (PAEC)

ROMANIA

Sumanariu C. A.	National Commission for Nuclear Activities Control

RUSSIAN FEDERATION

Bakhin A.	FSUE SRI SIA "LUCH"
Raspopov E.	TVEL JSC
Solntsev V.	FSUE SRI SIA "LUCH"

SWEDEN

Tronsén A.	Vattenfall Nuclear Fuel AB

SWITZERLAND

Zimmerman L.	Gösgen Nuclear Power Plant

TÜRKIYE

Ergin C.	Ministry of Energy and Natural Resources of Turkey

UKRAINE

Godun O.	Scientific and technical Centre of SE "National Nuclear Energy Generation Company "ENERGOATOM"

UNITED ARAB EMIRATES

Al Seyabi M.	Emirates Nuclear Energy Corporation (ENEC)
Alshehhi A.	Emirates Nuclear Energy Corporation (ENEC)

LIST OF PARTICIPANTS FOR 2023 TECHNICAL MEETING

ARGENTINA

Parrado M. F.	Comisión Nacional de Energía Atómica (CNEA)

ARMENIA

Arakelyan A.	Haykakan Atomayin Electrakayan CJSC
Petrosyan V.	Armenian Scientific Research Institute for NPP Operation - ARMATOM

BRAZIL

Alves de Mattos M.	Indústrias Nucleares do Brasil (INB)
Meireles J. B.	Indústrias Nucleares do Brasil (INB)
Oliveira de Souza P.	Indústrias Nucleares do Brasil (INB)
Paneto L.	Indústrias Nucleares do Brasil (INB)
Rodrigues Carnaval J. P.	Indústrias Nucleares do Brasil (INB)

BULGARIA

Ivanov N.	Kozloduy Nuclear Power Plant
Palakarcheva-Ivanova R.	Kozloduy Nuclear Power Plant
Pimpas A.	Kozloduy Nuclear Power Plant

CANADA

Dimayuga F.	Canadian Nuclear Laboratories (CNL)

CHINA

Fan H.	Nuclear Power Institute of China (NPIC)
Hong Z.	Nuclear Power Institute of China (NPIC)
Xu C.	Nuclear Power Institute of China (NPIC)

FRANCE

Bischoff J.	Framatome
Gribot Perrin N.	Framatome
Marie A.	Framatome
Mazaudier F.	Commissariat à l'Énergie Atomique et aux Énergies Alternatives (CEA)
Raveu G.	Framatome
Robisson A-C.	Commissariat à l'Énergie Atomique et aux Énergies Alternatives (CEA)

INDIA

Kalra H.	Nuclear Power Corporation of India Ltd.
Kumar P.	Bhabha Atomic Research Centre Mumbai

IRAN, ISLAMIC REPUBLIC OF

Ahmadi Olounabadi S. A.	Atomic Energy Organization of Iran (AEOI)
Amini S.	Atomic Energy Organization of Iran (AEOI)
Amirian Molkemian R.	Atomic Energy Organization of Iran (AEOI)
Mokhtari A.	Atomic Energy Organization of Iran (AEOI)
Pasandi A.	Atomic Energy Organization of Iran (AEOI)

KOREA, REPUBLIC OF

Kim D-J.	Korea Atomic Energy Research Institute (KAERI)
Kim H. G.	Korea Atomic Energy Research Institute (KAERI)
Lee J-H.	KEPCO Nuclear Fuel
Yang J. H.	Korea Atomic Energy Research Institute (KAERI)
Yang S-C.	KEPCO Nuclear Fuel

PAKISTAN

Siddique M.	Pakistan Atomic Energy Commission (PAEC)

ROMANIA

Dumitrescu I. M.	RATEN ICN
Nicolae Tiberiu C.	Institute for Nuclear Research
Popa K.	

RUSSIAN FEDERATION

Gusev A.	TVEL JSC
Liashko I.	Scientific and Engineering Centre for Nuclear and Radiation Safety
Radostin A.	TVEL JSC

SOUTH AFRICA

Tshivhombela K.	The South African Nuclear Energy Corporation SOC Limited

SPAIN

Lafuente Sevillano J. J.	ENUSA Industrias Avanzadas
Verdejo Garrido D.	ENUSA Industrias Avanzadas

UNITED ARAB EMIRATES

Al Mesaybeh M.	Emirates Nuclear Energy Corporation (ENEC)
Alshehhi A. S.	Emirates Nuclear Energy Corporation (ENEC)
Alteneiji S.	Emirates Nuclear Energy Corporation (ENEC)

CONTRIBUTORS TO DRAFTING AND REVIEW

Ablitzer, C.	Commissariat à l'Énergie Atomique et aux Énergies Alternatives, France
Ahufinger Muñoz, A.	ENUSA Industrias Avanzadas, Spain
Alvarez, L.	Comisión Nacional de Energía Atómica, Argentina
An, N.	International Atomic Energy Agency
Bae, Y.	KEPCO Nuclear Fuel, Republic of Korea
Bischoff, J.	Framatome, France
Blanc, N.	Commissariat à l'Énergie Atomique et aux Énergies Alternatives, France
Bussolini, A.	Comisión Nacional de Energía Atómica, Argentina
Carnaval, J. P. R.	Industrias Nucleares do Brasil, Brazil
Castaño Marcos, J.	ENUSA Industrias Avanzadas, Spain
Collin De L'Hortet, A.	Framatome, France
Dai, X.	Nuclear Power Institute of China, China
Dimayuga, F.	Canadian National Laboratories, Canada
Duguay, C.	Commissariat à l'Énergie Atomique et aux Énergies Alternatives, France
Fan, D.	China North Nuclear Fuel Corp., China
Ghatreh, R.	Atomic Energy Organization of Iran (AEOI), Iran
Gufron, H.	National Research and Innovation Agency, Indonesia
Iltis, X.	Commissariat à l'Énergie Atomique et aux Énergies Alternatives, France
Isnaini, A.	National Research and Innovation Agency, Indonesia
Jo, H.	KEPCO Nuclear Fuel, Republic of Korea
Kandeepan, K.	Framatome, France
Kerboul, G.	Orano, France
Kim, D-J.	Korea Atomic Energy Research Institute, Republic of Korea
Kim, D-S.	Korea Atomic Energy Research Institute, Republic of Korea

Kim, H. G.	Korea Atomic Energy Research Institute, Republic of Korea
Koo, Y-H.	Korea Atomic Energy Research Institute, Republic of Korea
Kultayev, Y.	International Atomic Energy Agency
Kwak, D.	KEPCO Nuclear Fuel, Republic of Korea
Lafuente Sevillano, J.	ENUSA Industrias Avanzadas, Spain
Langenati, R.	National Research and Innovation Agency, Indonesia
Lee, B.	KEPCO Nuclear Fuel, Republic of Korea
Liang, X.	China North Nuclear Fuel Corp., China
Marie, A.	Framatome, France
Marin, J-M.	Orano, France
Medina, J. P.	Comisión Nacional de Energía Atómica, Argentina
Mi, W.	China North Nuclear Fuel Corp., China
Minetti, A.	Comisión Nacional de Energía Atómica, Argentina
Namy, P.	SIMTEC, France
Nazem, H.	Atomic Energy Organization of Iran, Iran
Nicollet, C.	Framatome, France
Olivera Muñoz, M.	Comisión Nacional de Energía Atómica, Argentina
Parrado, M. F.	Comisión Nacional de Energía Atómica, Argentina
Qiu, J.	Nuclear Power Institute of China, China
Radostin, A.	TVEL, Russian Federation
Rahmadi, G.	National Research and Innovation Agency, Indonesia
Raveu, G.	Framatome, France
Riahi, A.	Atomic Energy Organization of Iran, Iran
Robisson, A.-C.	Commissariat à l'Énergie Atomique et aux Énergies Alternatives, France
Sim, K.	Consultant, Canada
Souza, P. O.	Indústrias Nucleares do Brasil, Brazil
Suryaman, G. K.	National Research and Innovation Agency, Indonesia

Vankabo, P.	National Research and Innovation Agency, Indonesia
Verdejo Garrido, D.	ENUSA Industrias Avanzadas, Spain
Viry, F.	SIMTEC, France
Xi, L.	China North Nuclear Fuel Corp., China
Xiao, L.	Canadian Nuclear Laboratories, Canada
Xu, C.	Nuclear Power Institute of China, China
Yang, J. H.	Korea Atomic Energy Research Institute, Republic of Korea
Yang, S.	KEPCO Nuclear Fuel, France
Yang, Z.	Nuclear Power Institute of China, China
Zahraie, M.	Atomic Energy Organization of Iran, Iran
Zhang, H.	China North Nuclear Fuel Corp., China
Zhao, W.	Nuclear Power Institute of China, China

Technical Meetings

Virtual Meeting: 8–10 November 2021, Vienna, Austria: 26–28 June 2023

Consultancy Meetings

Vienna, Austria: 15–17 April 2024

IAEA
International Atomic Energy Agency

CONTACT IAEA PUBLISHING

Feedback on IAEA publications may be given via the on-line form available at:
www.iaea.org/publications/feedback

This form may also be used to report safety issues or environmental queries concerning IAEA publications.

Alternatively, contact IAEA Publishing:

Publishing Section
International Atomic Energy Agency
Vienna International Centre, PO Box 100, 1400 Vienna, Austria
Telephone: +43 1 2600 22529 or 22530
Email: sales.publications@iaea.org
www.iaea.org/publications

Priced and unpriced IAEA publications may be ordered directly from the IAEA.

ORDERING LOCALLY

Priced IAEA publications may be purchased from regional distributors and from major local booksellers.

25-02803E

FSC
www.fsc.org
MIX
Papier | Fördert gute Waldnutzung
FSC® C083411

Zeitfracht Medien GmbH
Ferdinand-Jühlke-Straße 7
99095 Erfurt, Deutschland
produktsicherheit@kolibri360.de